Climate, Food and Violence:

Understanding the Connections, Exploring Responses

Quaker Institute for the Future Pamphlet Series

1—*Fueling our Future: A Dialogue about Technology, Ethics, Public Policy, and Remedial Action*, coordinated by Ed Dreby and Keith Helmuth, edited by Judy Lumb, 2009.

2—*How on Earth Do We Live Now? Natural Capital, Deep Ecology, and the Commons*, by David Ciscel, Barbara Day, Keith Helmuth, Sandra Lewis, and Judy Lumb, 2011.

3—*Genetically Modified Crops: Promises, Perils, and the Need for Public Policy*, by Anne Mitchell, with Pinayur Rajagopal, Keith Helmuth, and Susan Holtz, 2011.

4—*How Does Societal Transformation Happen? Values Development, Collective Wisdom, and Decision Making for the Common Good*, by Leonard Joy, 2011.

5—*It's the Economy, Friends: Understanding the Growth Dilemma*, edited by Ed Dreby, Keith Helmuth, and Margaret Mansfield, 2012.

6—*Beyond the Growth Dilemma: Toward an Ecologically Integrated Economy* edited by Ed Dreby and Judy Lumb, 2012.

7—A *Quaker Approach to Research: Collaborative Practice and Communal Discernment*, by Gray Cox with Charles Blanchard, Geoff Garver, Keith Helmuth, Leonard Joy, Judy Lumb, and Sara Wolcott, 2014.

8—*Climate, Food and Violence: Understanding the Connections, Exploring Responses*, by Judy Lumb, Phil Emmi, Mary Gilbert, Laura Holliday, Leonard Joy, and Shelley Tanenbaum, 2014.

Climate, Food and Violence:
Understanding the Connections, Exploring Responses

Judy Lumb, Phil Emmi, Mary Gilbert,
Laura Holliday, Leonard Joy, and Shelley Tanenbaum

Quaker Institute for the Future Pamphlet 8
Quaker Institute for the Future 2014

Contact the Quaker Institute for the Future <quakerinstitute.org> regarding any editorial matters. Feel free to quote, but please give proper credit: Lumb, Judy, *et al., Climate, Food and Violence: Understanding the Connections, Exploring the Responses.* Quaker Institute for the Future Pamphlet # 8.

Published for Quaker Institute for the Future by *Producciones de la Hamaca*, Caye Caulker, Belize <producciones-hamaca.com>
ISBN: 978-976-8142-59-7

Climate, Food and Violence: Understanding the Connections, Exploring Responses is the eighth in the series of Quaker Institute for the Future Pamphlets: Series ISBN: 978-976-8142-21-4

Quaker Institute for the Future Pamphlets aim to provide critical information and understanding born of careful discernment on social, economic, and ecological realities, inspired by the testimonies and values of the Religious Society of Friends (Quakers). We live in a time when social and ecological issues are converging toward catastrophic breakdown. Human adaptation to social, economic and planetary realities must be re-thought and re-designed. ***Quaker Institute for the Future Pamphlets*** are dedicated to this calling based on a spiritual and ethical commitment to "right relationship" with Earth's whole commonwealth of life.

Quaker Institute for the Future
<quakerinstitute.org>

Producciones de la Hamaca is dedicated to:

—Celebration and documentation of Earth and all her inhabitants,
—Restoration and conservation of Earth's natural resources,
—Creative expression of the sacredness of Earth and Spirit.

Contents

List of Sidebars

Abbreviations

AFSC	American Friends Service Committee
^{0}C	Degrees Centigrade
C_3	Three-Carbon (plant)
C_4	Four-Carbon (plant)
CAC	Conversatorios of Citizen Action
CBD	Convention on Biological Diversity
CFC	Chorofluorocarbons
CFSC	Canadian Friends Service Committee
CWFS	Committee on World Food Security
CH_4	Methane
CO_2	Carbon dioxide
CSD	Commission on Sustainable Development
ESCR	International Convenant on Economic Social and Cultural Rights
^{0}F	Degrees Farhenheit

FAO	UN Food and Agriculture Organization
FCNL	Friends Committee on National Legislation
FDA	U.S. Food and Drug Administration
FPIC	Free, Prior, and Informed Consent
FSMA	U.S. Food Safety Modernization Act
FTA	Free or Bilateral Trade Agreement
GHG	Greenhouse gases
HCHR	UN High Commissioner for Human Rights
ILO	International Labour Organization
IPCC	Intergovernmental Panel on Climate Change
KXL	Keystone Pipeline
MSP	Multi-Stakeholder Platforms
NGO	Non-governmental Organization
NO	Nitrous oxide
OCHA	Office for the Coordination of Humanitarian Affairs
PGR	Plant Genetic Resources
QEW	Quaker Earthcare Witness (North America)
QIAP	Quaker International Affairs Programme (Canada)
QIF	Quaker Institute for the Future (North America)
QPSW	Quaker Peace and Social Witness (Britain Yearly Meeting)
QUNO	Quaker United Nations Office (New York and Geneva)
RSWR	Right Sharing of World Resources
RWE	Rheinisch-Westfälisches Elektrizitätswerk (German Utility Corporation)
SNAP	Supplemental Nutrition Assistance Program
TPP	Trans-Pacific Pact Partnership
UN	United Nations
UNDRIP	UN Declaration on the Rights of Indigenous Peoples
UNFCCC	UN Framework Convention on Climate Change
UNHCR	UN High Commission on Refugees
UNICEF	UN International Emergency Children's Fund
U.K.	United Kingdom
U.S.	United States
WFP	World Food Programme

Preface

This pamphlet was prepared by a Circle of Discernment under the auspices of Quaker Institute for the Future[1] to elucidate current and future effects of climate change on food throughout the world, and to help create an understanding of how violent conflict over food-related resources can be prevented. A Circle of Discernment is a small research group that utilizes the collaborative and Spirit-led practices of the Religious Society of Friends (Quakers) to address current issues of human and planetary well-being.

Six Friends have met for over two years via conference call to raise concerns about potentially disastrous climate change impacts on our food systems and to highlight the most compelling areas where change is needed. Our concern is the minimization of global hunger and the prevention of deadly violence. Our focus is on the interactions, the nexus of climate change, food, and violence.

We examine these interactions accepting the conclusions of the mainstream scientific community that climate change is real and caused by increased emissions of greenhouse gases (GHG), mostly from the use of fossil fuels for various human activities since the dawn of the industrial revolution. Most GHGs have a long retention time in the atmosphere, which means that what has already been emitted will increase the global temperature by 2.9°F. This will be more than is needed to inundate several island nations and severely impact highly populated coastlines. We are already seeing the results of rising temperatures in the melting of the Arctic ice cap, increasing incidence of super-storms like Sandy and serious droughts like the one in California in 2014.

We use the more general term "climate change" rather than "global warming" because, while the changes in our climate are caused by the warming of the planet due mostly to human activities, the effects on food involve issues other than temperature, such as water availability and atmospheric carbon dioxide (CO_2). Rather than concern ourselves with projections to the end of this century, which are dependent on predictions of future GHG emissions, we consider issues of the next four decades. During this period human settlements and ecosystems will be experiencing changes wrought by the GHGs that have already been emitted as governments and civil societies attempt to feed an increasing human population.

We tell the story of how climate change is likely to impact food systems and how a response to this situation can be created that helps to forestall social unrest and deadly conflict. The actors in this story are those that have roles to play in the response to increasing hunger: farmers, consumers, communities, governments, international agencies, corporations, scientists, market systems, and financial speculators.

We explore what actions are needed to adapt to climate change, especially to minimize the experience of hunger and violence arising from food insecurity. It is the purpose of this pamphlet to provide awareness, meaningful perspective, and information in a concise, accessible form to help empower civil society and government policy makers to meet the challenge of feeding Earth's human population while preventing destructive and deadly conflict.

The proposal for this research project and QIF Pamphlet was originally brought to the 2011 Annual Meeting of Friends Committee on National Legislation (FCNL) in Washington. It was prompted by FCNL's "Smart Security" program, which had developed out of its decade long "War Is Not the Answer" education campaign. When QIF was established in 2003, its founding document described one of its roles as doing background research that would be helpful to frontline Quaker organizations such as FCNL and American Friends Service Committee (AFSC). This project was seen as building on a briefing paper FCNL had previously produced. A QIF Circle of Discernment was established shortly thereafter and this pamphlet is the result.

We are not alone in this work. We are grateful for input from reviewers, including Jose Aguto, Molly Anderson, Charles Blanchard, Alan Connor, Barbara Day, Lynn Finnegan, Geoffrey Garver, Frank Granshaw, Pamela Haines, Keith Helmuth, John Lodenkamper, Piniayur Rajagopal, Laura Rediehs, Michael Shank, and Sara Wolcott. They have shared their insights from their wisdom and extensive experience, all of which has contributed tremendously to this pamphlet.

Phil Emmi, Mary Gilbert, Laura Holliday,
Leonard Joy, Judy Lumb, and Shelley Tanenbaum

September, 2014

Introduction

David Gee says: "Food. In every bite is the sun's light, Earth's evolution, the labor of other people, and particles your body will soon make its own. In food culminate all the things that make life and communion possible—cosmos, Earth, people—it really is something to say Grace for. If we want to evaluate how well we love ourselves, others and the Earth, there is no better barometer than how we grow and eat food. With food, fundamental questions about how we live our lives and organize our societies are literally in your face."[2]

Governments and leaders of civil society in many parts of the world have the double challenge of feeding a growing human population in the face of major disruptions due to climate change. In both town and country, there are those who suffer hunger and malnutrition. In urban areas especially, rising food prices have caused social tensions, sometimes leading to deadly violence. Continuing climatic disruptions, land erosion, drought, flooding, and migration will aggravate existing conflicts. The global context of extensive poverty, population growth, and loss of livelihood leading to major migrations, together with weak and corrupt governments, and failing economic and governance institutions, suggests that the scene is set for increasing civil unrest and deadly conflict.

Hunger

Hunger has a long history. Famine is recorded in the Bible and was doubtless experienced in prehistoric times. All continents have documented histories of famines which have led to a high percentage of deaths in indigenous and tribal populations. Imprinted on our Western memory is the nineteenth century Irish famine when at least a million died and another two million emigrated. More recently an estimated 900,000 people died in the Ethiopian famine of 1984. The Live Aid concerts raised tens of millions in British pounds to save many more. These are but a few of the many incidents of drought, famine, hunger, starvation, and death that have occurred on all continents of the world and persist into our time. While naturally occurring conditions in the form of drought, pest, and disease, have played a role in famine and hunger, human failure to respond to the threat of hunger or mitigate its impact is what ultimately leads to famine. And it is mostly the poor who die.

The UN Food and Agriculture Organization (FAO) estimated that 870 million of the 7.1 billion people on Earth were undernourished in 2012. Most (852 million) live in developing countries, where they represent 15 percent of those populations. In the past twenty years the percentage of undernourished people has decreased in Asia, the Pacific, Latin American, and the Caribbean. But in Africa increased food production has not kept pace with population growth. The proportion of people experiencing hunger in Africa has increased to nearly 25 percent over the same period.[3]

Since 1960, with only an eight percent increase in the amount of land used, the global food supply has increased by 170 percent due to improvements in agricultural methods. This has more than kept up with human population increases. The current global average of cereal production is one kilogram per person per day, which is plenty to feed the human population on Earth today. However, the FAO says that food production will have to be increased by 70 percent by 2050 to keep up with the growing population and changing food preferences.[4]

The application of new technology can lead to increased hunger, even as it increases the food supply. The "Green Revolution" began after World War II out of a fear that if food production were not increased, we would not be able to feed the growing human population. Early experiments in Mexico in 1948 improved the yields of maize and wheat. Varieties of seeds were introduced so that plants photosynthesized faster and devoted more energy to grain production than to leaves and stalks. But they needed more water, chemical fertilizers, and pesticides. By the 1960s the Green Revolution also included rice and potatoes and had spread throughout Latin America and south Asia. In five years India and Pakistan doubled their food production.[5]

While the Green Revolution did increase production, in other ways it was a disaster. From 1970 to 1990 food availability per capita worldwide increased, but the number of hungry people also increased. For example, it turned India from a grain importing country to a grain exporting country with increased hunger and malnutrition. New technologies, especially mechanization, led to the displacement of small-scale farmers. Only rich farmers in India could afford the required chemicals and the deep bore holes to draw ground water for irrigation. All of this led to lowering water tables, soil erosion, soil degradation, ground water contamination, and increased air pollution. It increased wealth for some and deepened poverty for many, with concomitant landlessness, and migration to urban areas. In response, urban poor were fed by dumping inexpensive food that is grown in places where farmers are subsidized to produce food at below production cost with high use of fossil fuel-based inputs. This supported urban populations at the expense of rural populations in India.[5]

Even with all these changes, small-scale farmers still produce most of the food that feeds the world's population, so future food availability depends on small-scale farmers having access to land with adequate water and soil nutrients. Irrigation was used in 2013 on 17 percent of croplands that produce 40 percent of the world's food. More irrigation will be needed as some areas experience decreased rain, snow, and glacier-melt water.

But ensuring that there is more than enough food for everyone does not ensure that everyone is fed. Inadequacy of overall food production has not been the cause of hunger and malnutrition. Food can be produced, but access to it is what matters. One must have either the land to produce one's own food, the funds to purchase this basic necessity, or social networks and safety nets that help to share the available food.

Economic factors play a large role in food distribution. To resolve hunger, political and civic leaders must address inequality, both within countries and globally. Hunger in the most vulnerable populations is a critical manifestation of poverty, of lack of livelihood. One out of seven persons in the U.S. received assistance from the Supplemental Nutrition Assistance Program (SNAP) in 2011.[6] They would have gone hungry were it not for this assistance, but not because there was a shortage of food. As this pamphlet goes to press, the 2014 Farm Bill passed by Congress and signed by President Obama cut $8 billion from the total SNAP, which is about one percent of the total SNAP budget and affects about 850,000 households in the U.S.[7]

The biggest challenge in addressing hunger is to ensure sustainable livelihoods, both rural and urban. Poverty and hunger are interrelated because poor people do not have the resources to buy food or acquire land. Hunger can exacerbate poverty because poor health, low energy and mental impairment reduce people's ability to work. And poverty and hunger are frequently caused, and triggered, by violent conflict. Violence often results in internally displaced persons and international refugees. These problems would exist even if there were no climate change, but the changing climate exacerbates these problems and will continue to do so.

Climate change is likely to have significant effects on food systems: some positive, but most negative. The effects of climate change are likely to exacerbate already existing unequal distribution of wealth and natural resources that are at the root of food insecurity.

Using this Pamphlet

Global food production has been more than adequate to feed the global population, but growing population pressure, rising incomes with changes in food preferences, the competition for land and water—all in the context of climate change—make feeding the world an increasing challenge. This pamphlet is designed to facilitate discussion of this concern and to prompt appropriate action.

Chapter One explores the many ways in which climate change affects temperature and rainfall with implications for the incidence and severity of floods and drought, and creates critical uncertainties in the timing of water availability.

The implications of these changes on food production are considered in Chapter Two, including availability of arable land and water, effects of climate change on crop production, livestock production, pests and pathogens, and the contribution of food production to climate change.

But these are not the only forces affecting people's access to the food they need. Chapter Three explores economic, legal, and institutional factors that control access to land, water, seeds, and other farm inputs, as well as those that control the composition and disposition of the food produced.

Many factors are driving the world's evolving food system, from the rising swell of immigration to the threat of increasing hunger, consequent social unrest, and violence. The resulting potential for conflict and deadly violence is discussed in Chapter Four.

The many responses to the disruptions brought by climate change already developing are described in Chapter Five.

Finally, after setting out guiding principles, Chapter Six proposes specific actions toward ensuring food security and a more peaceful future.

CHAPTER ONE
Climate Change is Happening

"There was a flood and my house was surrounded by water. I could no longer go out and play because there was no dry ground. I am the present and the future, a victim of climate change. I live in an area that is constantly being affected by disasters. Hurricane and floods are my reality. My life is real and so am I. Please do not treat climate change as a separate problem. It is a part of reality and that is why I am here today."
Walter Perriot, 11 years old, Crooked Tree Village, Belize, Child Ambassador to COP-16, sponsored by UNICEF

The social impacts and economic disruptions of climate change are real. The scientific community has concluded that it is mostly caused by human activity.[8] The 2013-14 Intergovernmental Panel on Climate Change (IPCC) report states *"Warming of the climate system is unequivocal, and since the 1950s, many of the observed changes are unprecedented over decades to millennia. The atmosphere and ocean have warmed, the amounts of snow and ice have diminished, sea level has risen, and the concentrations of greenhouse gases have increased."*[9]

In their assessment of future threats based on extrapolation from existing trends, the U.S. National Intelligence Council states that by 2030, "demand for food, water, and energy will grow by approximately 35, 40, and 50 percent respectively owing to an increase in the global population and the consumption patterns of an expanding middle class. Climate change will worsen the outlook for the availability of these critical resources. Climate change analysis suggests that the severity of existing weather patterns will intensify, with wet areas getting wetter, and dry and arid areas becoming more so."[10]

Yes, climate change is happening. In the U.S., legislative and other responses at the federal level have been hampered by a concerted effort supported by the fossil fuel lobby to claim that climate change is an unproven controversial theory, rather than the consensus of virtually all scientists.[11] But the reality is that the careful science of the IPCC has consistently predicted the impacts of the changing climate using climate models designed to project the effects of increasing GHGs in Earth's atmosphere on Earth's climate. Coordinated model experiments are run using the same initial conditions under scenarios with different assumptions about the future, so they present a range of projections of future climatic conditions. Observations of GHG emissions since 2000, reported in the 2007 IPCC fourth assessment, followed the trajectory of the more pessimistic IPCC projections. But in the 2013-14 fifth assessment, observations since 2007 followed the central projections, indicating the beginning of a slowdown in GHG emissions worldwide. The next sections summarize specific impacts of climate change that are likely to affect food.[12]

The Earth is Getting Warmer

Earth receives energy from the sun and radiates it back to space as heat. Greenhouse gases (GHGs) cause warming because they absorb part of the re-radiated heat. The main GHGs that trap heat in Earth's atmosphere include carbon dioxide (CO_2), methane (CH_4), nitrous oxide (NO), and chlorofluorocarbons (CFCs). Without the additional GHGs emitted through human activity, these gases would be exchanged among the oceans and the atmosphere, maintaining steady concentrations of CO_2 and CH_4 in the atmosphere and a balance between solar radiation and re-radiated heat. However, the emissions associated with human activity have thrown that system out of balance. In 2013 the CO_2 concentration was 395 parts per million; by 2050 the CO_2 concentration is projected to be about 550 ppm. In contrast to most other GHGs that have defined lifetimes in the atmosphere, most CO_2 is not broken down chemically in the atmosphere. Since it is exchanged with CO_2 in the oceans and incorporated into plants via photosynthesis, there is no defined lifetime for CO_2 in the atmosphere.[13]

Climate Change is Increasing Extreme Events

The most dramatic evidence of climate change has been the increasing incidence of extreme events such as heat waves, droughts, fires, and floods.[14] A study of the summer temperature distribution compared the frequency of temperature highs in the Northern

Extreme Events in Houston

"I had no words and I could not pray. It was in this moment that I became the prayer I could not utter.

"In June, 2001, tropical storm Allison visited my city and floodwaters destroyed my home and business. I am a native to this area, living at the highest elevation in Houston I had never experienced flooding and the destructive consequences of a storm of this magnitude. 38.6 inches of rain fell over a six-day period and flooded 95,000 automobiles and 73,000 houses. It destroyed 2,755 homes, leaving 30,000 homeless with residential damages totaling 2.29 billion.

"On August 29, 2005, Hurricane Katrina hit the Gulf Coast of the United States. Although it did not directly strike my city, refugees from cities along the Gulf Coast flooded into the city. Three weeks later Hurricane Rita made landfall on the Gulf Coast causing 3 million people to flee. This was the largest evacuation in United States history. I gathered my family and pets and fled north. A trip to the hill country of Texas that normally takes 3 hours took 18 hours. I saw people dying on the highway and witnessed municipal and state governments totally unprepared for this event.

"On September 13, 2009, the worst hurricane in the history of my city struck: Hurricane Ike. My home and business were again wiped out. I sat in my boarded up house and listened to sustained winds in excess of 100 mph howl for six hours. I heard debris flying, hitting the walls and saw sections of my roof blow away. I saw water come underneath doors and through windows. I stood in lines for over eight hours waiting to get one bag of ice and was without power for three weeks. Another 2.8 to 4.5 million people were also without power for weeks and some for months. I don't know whether all this devastation is due to climate change. All I know is that storms are more frequent and more intense. Prior to this time, I had read about events like these but never experienced them. I had consistently donated to the Red Cross and served as a volunteer to help others. This time, I was on the opposite end of need. These storms are real."

—Laura Ward Holliday

Hemisphere averaged over 1951-1980 with those of subsequent decades, demonstrating the increase in temperature extremes.

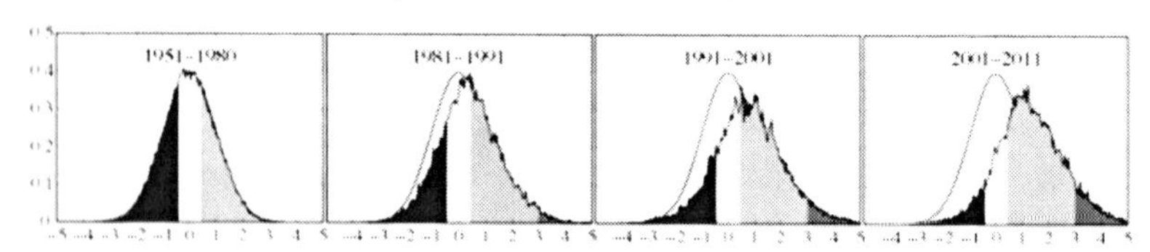

This shows that the average temperature has moved to the right toward higher temperatures. The black is cool; white is normal, the average of 1951-1980; light gray is warm; and darker gray is extremely hot. The curve flattens as the frequency of extremely high temperatures increases.[15]

One extreme event was the excessive rain that fell in 2013 in Colorado—15 inches in a two-day period. That was a 1,000-year event, that is, the probability it would happen in any one year is 0.1 percent.[16]

Not all extreme events associated with climate change result in local warming. In the winter of 2012-13, the arctic sea ice melted to a greater extent than ever recorded. This caused the jet stream to weaken and loop down into the eastern half of the U.S, which brought cold arctic air and a long, cold winter to the U.S. Midwest, while Alaska and Greenland had record high temperatures. A similar jet stream condition occurred in the winter of 2013-14 when the United States and the United Kingdom experienced cold and increased precipitation, while Scandinavia and Alaska were much warmer than normal.[17]

Rainfall

Earth's water cycle is powered by the energy from the sun, so warming will greatly affect rainfall. Increased global temperature will increase evaporation and potentially more rainfall in areas with low enough temperature to generate condensation. That is why high latitudes are getting increased rainfall, and the mid and low latitudes are getting less rainfall. The tropics are projected to get drier and semi-arid tropics to become more arid. More frequent and more severe floods and droughts are likely to affect larger areas of the world. Droughts and flooding may happen in the same areas, but in different seasons.[18]

Seasonal rainfall patterns are already being disturbed. Monsoon patterns will be increasingly disrupted, bringing uncertainties about the timing of seasonal rainfall onset. Time of planting is important for most non-perennial crops. It is geared to both soil moisture and temperature, in many areas critically so. Responses to new uncertainties about timing of rainfall include choosing crops that are less susceptible and spreading time of planting over the period that might cover the optimal planting date. But the consequence of this could be to reduce overall yield. The more unreliable the rainfall, the greater the likely reduction of yield and output of some crops, especially grains.[18]

As early as 2020, many African countries are projected to have such increased water stress that the yield from rain-fed crops will be half of what it is today. In Asia freshwater is projected to be decreased in large river basins, and delta regions are projected to have increased flooding from the sea and rivers. Agricultural production in southern and eastern Australia is projected to decrease because of drought and wild-fires. In Latin America, increased glacial melt is already causing floods, but is projected to cause desertification as glaciers recede. In some areas of the northern plains of North America where rain-fed crops are not at the upper level of their range, increases in productivity are predicted, but wherever crops are now at the upper temperature of their range, productivity is projected to decrease. In areas such as the U.S. Southwest that become drier, more groundwater will be needed for irrigation.[18]

Melting Glaciers and Ice Sheets

Mountainous regions that depend on summer glacier and snowpack melt for their water, such as the Andes and the Himalayas, are projected to become much drier as glaciers recede. Streams that depended on glacier melt will have decreased water flow. Over a billion people live downstream of those areas and depend on that water. The melting of glaciers is the most visible effect of climate change. From satellite data, a group of 47 scientists from 26 institutions demonstrated the decrease in the ice sheets of Greenland, West Antarctica, and the Antarctic Peninsula between 1992 and 2011.[19]

Even in those areas where annual precipitation may remain constant, stream flow will diminish during dry seasons as areas feeding those streams are de-glaciated and flooding will increase during wet seasons in these same areas as less of the winter precipitation is stored as snow. In both the Andes and Himalayas a very real concern is catastrophic flooding resulting from the failure of glacial lakes produced by receding glaciers.

Sea Level Rise

The global mean sea level has changed significantly throughout Earth's history. During the last glacial maximum about 20,000 years ago, when the average global temperature was 10°F colder, sea level was 350 feet lower than it is currently. In contrast, during the Pliocene three million years ago, the climate was 5°F warmer and the seas were 90 feet higher than today.[20] The current rise in global sea level is due to thermal expansion of warming ocean waters and the addition of melt water from ice sheets, glaciers, and mountainous ice caps.

The rate at which the sea level rises due to increasing temperature is slow at first, but then it accelerates and tapers off at a higher level, so the effect of global temperature increase on sea level has a significant delay. The sea level was relatively stable for the last 3,000 years, rising at a slow rate, but in the last 20 years that rate of increase has doubled. Projections put the rate of sea level rise at five times recent rates by 2100; so the greatest sea level rise from current warming is projected to happen in the 22nd century.[21] Current CO_2 levels have already locked in a four-foot sea level rise, which is enough to submerge more than half of the coastal U.S. cities at high tide.

Sea Level Rise on Caye Caulker, Belize

The island where I live will not survive to the end of this century. I live at about four feet above sea level; the highest point on the island is only six feet higher. High tides are increasingly flooding areas that never flooded before. Already rising sea level has caused so much erosion that sea walls have appeared all along the east side. About 1600 people live on this small island and many others live in other areas of Belize that are predicted to be inundated by 2100. There is no talk yet of out-migration, just adaptation for the near future.

—Judy Lumb

The effect of sea level rise at local and regional scales is exacerbated where the land is subsiding due to groundwater pumping or oil and gas extraction. Studies of river deltas and coastal cities have found that most deltas are sinking, and the risk of flooding in the cities due to storm surge, wind, tides and sea level rise has increased tremendously. Over the past decade, 85 percent of river deltas have experienced severe flooding inundating over 100,000 square miles of land, much of it prime agricultural land. There is now a 50 percent chance over five years that one or more coastal U.S. cities will be flooded by what was considered to be a one-in-1,000-year event.[22]

Economic Growth and Climate Change

The 1972 book, *The Limits to Growth*, showed various trajectories overshooting global resource limitations by the middle of the 21st Century. Their main message was that the planet has limited capacity to sustain economic growth and to exceed that capacity risks the collapse of industrial civilization.[23]

On the 40th anniversary of that ground-breaking book, three of the original authors assessed current reality in comparison to their prior forecasts and found that the trajectory of the past 40 years followed their scenario that exceeds the planet's capacity. If current population and economic growth trajectories continue, global temperatures will advance toward catastrophic levels, with an accelerating increase in extreme weather events and related disasters. A large share of capital investment will have to be diverted to adaptation and damage control, so less will remain for human needs and environmental restoration. There will be huge differences both within and among major regions. Resource mal-distribution will create more poverty.[24]

There is Still Time

The good news from the IPCC 2013-14 report is that there is still time to make necessary changes to avoid the most catastrophic impacts of climate change. The report provides much information on specific responses required, some of which will be discussed in Chapter Five. But the report emphasizes that the time to respond to climate change is now. The longer it takes to make the necessary, inevitable changes, the more it will cost, and the less likelihood there will be success.[25] A knowledgeable citizenry can better prepare and make the needed changes. It is the purpose of this pamphlet to contribute to that knowledge.

Summary: The Effects of Climate Change

Worldwide climate events such as hurricanes, tornadoes, droughts, and floods are occurring with more frequency and greater intensity. Other consequences of climate change include glaciers melting, sea level rising, surface temperature increasing, snow cover decreasing and higher humidity on a global scale. Some significant impacts of climate change are subject to positive feedback loops, which means they are both caused by, and contribute to increasing, climate change: a decrease in areas covered by ice and snow, which reflect sunlight, leads to more global warming. Low-lying island nations are threatened with extinction. For them the threat is one of survival as a country and as a people. For farmers worldwide, climate change will bring resource constraints, uncertainty, and major adaptive challenges that will result in lower crop outputs and food supplies. For many it will mean increasing food insecurity and the prospect of hunger. The next chapter describes current and potential effects of climate change on food production.

Environmental Disruptions Linked to Greenhouse Gas Emissions

Interactions between air and land: *Near-surface temperature increase. More intense droughts. Soil desiccation. Desertification. Dust storms. Increased heavy precipitation events.*

Interactions between air and ice: *Arctic ice cap shrinkage. Polar ice sheet melting. Glacial retreat. Permafrost thawing. Methane release.*

Interactions between air and water: *Oceanic thermal expansion. Increased atmospheric moisture. Higher energy storm cells. Ocean acidification. Coral bleaching.*

Interactions between water and ice: *Decreased arctic ice. Decreased reflection from arctic ice. Sea level rise. Weakened ocean currents. Glacier-fed streams flood and then dry up.*

Interactions between land and water: *Increased tidal surges, beach erosion, salination of coastal soils, and loss of wetlands. Increased storm floods, soil erosion, pollution of waterways, and oceanic dead zones.*

Interaction between ice and land: *More tundra fires, frost heaves, uneven thawing, glacial ice wedges, lakes and earth domes from melting and erosion of permafrost, landslides, shoreline damage from diminished sea-ice protection, unstable forests, and less mountain snowpack.*

Chapter Two

Food Production and Climate Change

Food production depends on the availability of land, sufficient water, seeds, pollinators, livestock, and control of weeds, pests and pathogens. The effect of climate change on food production is the focus of this chapter. The U.S. Department of Agriculture released a comprehensive report that delineates the effects of climate change on crops, livestock, soil, and water, and provides guidance for adaptation and future research.[26]

Climate Change and Arable Land

The land surface area of Earth is 36 billion acres, but only 8 billion acres is suitable for agriculture (arable land). Right now our human population is over 7 billion and growing by 211,000 persons per day with an expectation of reaching 8.5 billion by 2025.[27] This means that by 2025 there would be less than one acre of arable land per person, if the amount of arable land were to remain fixed.

However the amount of arable land is not fixed. It is being lost at the rate of 38,610 square miles (24.7 million acres) per year due to a combination of effects of climate change and human activity. Sea level rise, drought, desertification, deforestation, erosion by wind and water, soil degradation, poisoning by extractive industries, and the spread of cities all contribute to loss of arable land.[28]

Submersion of islands and lowland shore areas from rising sea levels, along with loss of land from the increase in extreme weather events, has already begun. The island nation of Tuvalu is slowly being submerged. The people of Tuvalu have requested permission to immigrate to New Zealand.[29] The islands in the Chesapeake Bay are being submerged.[30] Bangladesh's lowlands farmlands are being submerged and saltwater is seeping into inland fresh water systems.[31]

Desertification from overgrazing, over-exploitation for firewood, and salinization from incorrect use of irrigation leads to flooding, reduced water quality, dust storms, and pollution. Concern about increasing desertification, especially in Africa, was expressed in Agenda 21 at the 1992 Earth Summit in Rio, which led to the creation of the UN Convention to Combat Desertification.[32]

When forests are cut down to clear land for agriculture, or to provide wood for fuel or usable lumber, arable land acreage is increased, but the land can be degraded. Many government agencies are fighting illegal logging to protect their forests, but encouraging legal logging. However, whether legal or illegal, clear-cut logging leads to land degradation. Tree roots draw up water, which is then released into the atmosphere via transpiration. Trees trap rainwater, which allows the soil to absorb the flowing water. When there are no trees, water runs off, leaving less recharge of groundwater. This leads to reduction in groundwater resources and alternating periods of flood and drought in the affected areas.

Soil is degraded when forest areas are cleared. The soil is exposed to the sun, making it very dry and eventually, infertile, due to the loss of volatile nutrients such as nitrogen. Rain washes away other nutrients, which flow with the rainwater into waterways. Large tracts of land can be rendered permanently impoverished due to soil erosion and degradation.

Trees act as a major storage depot for carbon, since they absorb CO_2 from the atmosphere. When deforestation occurs, many of the trees are burnt, which releases the carbon that is stored in them as CO_2. Deforestation is a major contributor to the increasing greenhouse gases in the atmosphere that are causing global warming, climate change, and climate disruption.

By October 2011 when the global population was seven billion, about 24 percent of global land area had been affected by land degradation. Globally, 1.5 billion people—42 percent of the very poor—live on degraded lands.[33] Population is growing fastest in developing countries. In Sub-Saharan Africa, annual population growth of 2.3 percent has led to ever smaller areas of arable land per capita. Continuing land degradation exacerbates the effects of climate change.[34]

Climate Change and Water

Most global crop production still depends on rainfall as a water source. In the 1960s that proportion was 90 percent, but now it is around 80 percent, as most land being added to crop production requires irrigation.[35]

Water is available from atmospheric moisture falling as rain and snow, stream flow, and groundwater from aquifers. Irrigation for agriculture accounts for 70 percent of all groundwater extracted worldwide.[36] River-borne sources are dependent on upstream rainfall, snowfall, or ice melt. All of these sources are likely to be affected by climate change.

Much river-based irrigation is controlled by dams, commonly associated with electric power generation, the demands for which are often not compatible with farmer needs, a likely source of tension. Reservoirs created by dams collect silt at an average rate of one percent per year, unless effective soil erosion management is in place. In roughly 50 years, a reservoir is half full of silt and has lost its usefulness.[37]

Many rivers flow through several countries, requiring international agreements. For example, the Himalayan watershed provides water in several rivers to seven countries, many with dense human populations and high concentrations of poverty. From the western slopes of the Himalayan Mountains, climate change will likely bring significant reduction of river flows as the glaciers recede and provide less and less melt-water to India and Pakistan in the Indus River, exacerbating tensions between those two countries. In Bangladesh on the eastern slopes of the Himalayan Mountains, the problem is flooding. All this is complicated by China building many dams on the Tibetan Plateau, making Vietnam, Laos, Thailand, and Cambodia nervous because, while these countries have mechanisms in place to negotiate water issues, so far China has refused to participate in intergovernment dialogues.[38]

Using groundwater for irrigation is dependent on recharge of the aquifer. Currently, there are several areas where recharge is failing to maintain water levels in deep aquifers, potentially affecting millions of farmers. For example, Egypt has two sources of water, the Nile River and groundwater under the desert. Both are decreasing because recharge is not happening as fast as usage, and some of the water is too polluted for use. Egypt has a major challenge to provide clean freshwater for the needs of its growing population, for irrigation to produce food and for sustainable development.[39]

The Ogallala aquifer, which underlies a large agriculture area in the United States plains, is being recharged at only 15 percent of what is being used. At that rate it will be depleted by 2050.[40]

Even where water is not scarce, it is not always readily available to all users. For example, Dalits, a lower untouchable caste in India, are not allowed to draw potable water from government-dug deep tube wells.[41] And they are last in line for surface water irrigation. Inequitable distribution of water leads to increasing distress and tension.

Vulnerability of Food Crops to Climate Change

The vulnerability of food crops to climate change depends on the sensitivity of individual crops to temperature and water, and whether they are being grown at conditions close to their upper tolerance levels.[42]

In areas where plants are growing at temperatures significantly lower than their lethal levels, increasing temperatures usually increase the yield because they extend the length of the growing season. As growing conditions approach plants' lethal upper temperature limit, yields decrease. Some increase in temperature can be tolerated if adequate water is available, but that can induce early flowering or other changes in growth patterns. Plant reproductive organs are generally more sensitive to temperature increases and water scarcity; so the timing of rainfall is another factor that determines yield.

The response of particular plants to the effects of climate change—water stress, increasing temperatures, and increasing atmospheric CO_2—depends on their pattern of photosynthesis. Plants have different patterns of photosynthesis called "C_3" and "C_4" which is the number of carbon atoms in the initial molecule formed by addition of CO_2.

C_3 plants have a simple system involving stomata in the leaf surface that allow CO_2 through where it is incorporated directly into a three-carbon molecule in the Calvin-Benson cycle for photosynthesis. The open stomata also allow water to evaporate in a process called "transpiration" which cools the plant.

C_3 plants thrive in conditions of moderate sunlight and temperature with plenty of soil moisture because through transpiration they lose 97 percent of the water taken up by their roots. Under hot temperatures C_3 plants close their stomata to stop water loss, which also reduces uptake of CO_2, and consequently reduces the plant's growth.

C_4 plants have an additional layer of cells and an additional pathway in which CO_2 is incorporated into a four-carbon molecule, and then transferred to the inner layer of cells where it enters the Calvin-Benson cycle in a more efficient chemical process. Because this reaction occurs away from the stomata, there is much less water loss. Therefore, C_4 plants are more resistant to water stress than C_3 plants.

Most of the 86 crops that feed humanity on Earth are C_3 plants, including the major crops of wheat, rice, and soybeans, as well as nearly all cereals, legumes, fruits, roots, and tubers. Rice and wheat, being C_3 plants, show an increase in yield when growing under higher CO_2 concentrations, but they are less nutritious because it is only carbon that increases, not nitrogen, which results in lower protein levels. Soybeans are legumes, which fix nitrogen from the atmosphere, so they show very little decrease in protein content. Rice, wheat and soybeans all show a decrease in mineral content, which has a serious effect on human nutrition because already a large part of the global population is deficient in iron, zinc, and vitamin A. Cereal crops are often used as feed for livestock, so the resulting meat from those livestock is also deficient in mineral content.

There are only four major C_4 food crops—maize, sorghum, millet, and sugarcane—but these four comprise 20 percent of total food consumption. They are more temperature and drought-resistant than other food crops. Two-thirds of maize production is in developing countries, especially Latin America and Africa. The demand for maize will double by 2050, but climate change may reduce the yields.[43]

Production of Food Crops in Light of Climate Change

Crop	*Characteristics*	*Total*[a]	*Effect of Climate Change*	*Effect in U.S.*
maize	C4 high yield per hectare, drought resistance	819	Yield increases, then drops quickly over 29°C	Losses of 43—79% by 2050
wheat	C3 moderate temperature, plenty of water	686	Lose10% for every 1°C rise over 34°C 10-14% decrease in protein with increased CO_2	
rice	C3 both tropical and temperate varieties, plenty of water	685	Lose10% for every 1°C rise over minimum night temperature,10% decrease in protein with increased CO_2	
potatoes	C3 suited to cool temperatures	330	Sensitive to temperature increase, 10% decrease in protein with increased CO_2	
cassava	C3 well adapted to drought and heat, tolerant to lowfertility	218	Yield increases, high optimum temperature and positive CO_2 response	
soybeans	C3 legume	212	Yield increases, then drops quickly over 30°C, Only 1.5% decrease in protein under increased CO_2	Area suitable drops by 60% by 2070; Lose 30—82%
barley and rye	C3 high tolerance to salinity, drought	170	10-14% decrease in protein with increased CO_2	
yams and sweet potatoes	C3 well adapted to drought and heat	152		
millet and sorghum	C4 hardy, wide temperature range, grow in poor soil	83	Yield decreases by 2.5% by 2050	

[a] *Total Global Production in 2009 (mega-tons)*[43]

Climate Change, Pests, and Pathogens

The destructive effects of pests and pathogens are an important determinant of food crop productivity. The potential effects of climate change on insects, bacteria, and other pests are complex. Increasing CO_2 concentration, increasing temperature, and drought or increased water availability all must be taken into account. The considerable capacity of insects and bacteria for rapid genetic mutation means that they are likely to take advantage of the changing conditions wrought by climate change. Higher temperatures might allow their geographical range to be expanded, the timing of their appearance to be earlier in the growing season, and their numbers increased, all of which can have a deleterious effect on crop yield. Some effects of climate change on pathogens are already obvious.[44] For example, more frequent earlier infections of potato late blight have been observed in Finland over the past 70 years.[45]

Climate Change and Livestock

More than 600 million people worldwide make their living raising livestock. Water and feed are the most significant inputs into this industry; so the effects of climate change on food crops are also reflected on livestock production, including feed crop and water availability, quantity and quality of rain-fed forage for pasture, and effects on pests and pathogens. Heat stress has significant effects on the animals. Higher temperatures cause feed intake, conception rates, and milk production all to decrease. Recent extreme summer temperatures have caused significant livestock mortality in Europe and the U.S.[46]

Global warming of 2^0C is likely to increase pasture livestock production in humid regions, but decrease production in semi-arid or arid areas. Rangelands are complex ecosystems with the plant composition varying even between winter and summer. Increasing temperature favors C_4 plants because of their increased resistance to drought, but increasing CO_2 concentration favors C_3 plants. Increased yield in C_3 plants due to higher incorporation of carbon is countered by decreased nutritional value, because there is no concomitant increase in nitrogen except in legumes that fix nitrogen from the atmosphere. This effect may be balanced because the proportion of legumes in grasslands is increasing. Under temperature and water stress, plants tend to increase lignification, or the "woodiness" of their stems, which decreases digestibility and forage quality of the rangelands.[47]

In tropical and sub-tropical areas, the effect of heat stress is expected to be less because livestock breeds raised in those areas are already well adapted to heat and drought. But at higher latitudes, livestock breeds may be currently located where summer temperatures will exceed their maximum comfort zones. As with plants, livestock production might increase at higher latitudes, and genetic varieties developed for lower latitudes may be more appropriate for higher latitudes as they become warmer.

Confined animal facilities have more control over environmental extremes, but create major air and water pollution. Windbreaks can be used to reduce the effect of winter storms, and shade and sprinklers can be used to reduce the effect of heat waves in the summer. However when many animals are concentrated in small areas, they are very vulnerable to pathogens, so antibiotics are given prophylactically. This routine use of antibiotics selects for pathogenic bacteria that are resistant to antibiotics, as in the appearance of "super-bugs" that are resistant to many antibiotics. This is an increasing problem in both animal and human health,

Climate Change and Fisheries

Fish and other aquatic food products provide 20 percent of the protein in human diets worldwide, especially among the poorest in developing countries where fisheries directly employ and indirectly support half-a-billion people. In developing countries small-scale marine and freshwater capture of wild fish makes the biggest contribution to food security.[48]

In contrast to agricultural food production, aquatic food production involves mostly wild capture methods, but both capture fishing and aquaculture are important in feeding humans. In 2011, the global aquatic food production was 131 million metric tons, of which 64 million tons (44%) was from aquaculture. Marine capture fish production has remained relatively stable since 1985, which indicates that this resource is fully exploited. However, freshwater capture has doubled over the same time period.[49]

Increasing temperatures due to climate change are likely to increase capture fish production in higher latitudes and in high altitude deep lakes due to increases in growing seasons and reduction of ice cover, but toward the Equator productivity is likely to slow down.[50]

Like confined livestock production, genetic varieties have been selected for aquaculture, and growth conditions can be somewhat controlled, but the fish are highly susceptible to disease and cause major pollution. The ranges of diseases are increasing and moving into higher latitudes as temperatures increase. Many aquaculture facilities are located in delta areas, which will be subjected to saltwater intrusion as sea level rises. Some aquaculture operations utilize cage cultures that will be impacted by extreme weather situations.[51]

Contribution of the Food System to Climate Change

The current trend in industrialized agriculture of continually growing a single crop on large expanses of land began in the 1950s when the implied message to farmers from the U.S. Department of Agriculture was "get big or get out." This advice and the policies that accompanied it were designed to increase farm production through mechanization, and make large quantities of food available to commodity markets for trading and speculation. Large fields of single crops require chemical pesticides because they are easy

Family Farms Getting Bigger

I am a retired farmer, so I rent my land out to neighboring farmers. When I was farming this land, my planter inserted seeds in eight rows at a time and then I had to come back again to apply fertilizer. Current sophisticated planting equipment uses soil testing and GPS location devices to apply exactly the combination of fertilizers needed for soil conditions. And the modern planters cover as many as 36 rows while applying the fertilizer needed in that location in one pass through the field. The result is more efficient use of fossil fuel, but greater dependency upon it.[52]

It is not that family farms are disappearing. Family farms still exist, but there are fewer of them. Because these large pieces of equipment are expensive, fathers and sons or brothers have banded together to farm more and more land.

—Wallace McGee, Kincaid, Kansas

targets for pests. Continuously growing the same crop on a plot of land depletes the soil, so large amounts of chemical fertilizers must be used. The result is that most of U.S. agriculture is now heavily dependent on fossil fuels for production of chemical fertilizers and pesticides, to run the machinery and to bring agricultural products to market.

Many food crops, including livestock, have become commodities traded in national and world markets, enabling transnational corporations to accumulate large profits and control large sectors of the food economy. Under pressure to produce large profits, industrial agriculture today in the U.S. ignores considerations of environmental and social costs and fails to respect the ecological integrity of the soil and the wellbeing of rural towns and communities. There is a conflict of interest between maximization of short-term profit and safeguarding the health of the soil and human communities. Agricultural land has been consolidated in ever-larger corporate holdings with ever intensifying forms of mechanization, all tied to increasing use of fossil fuels. While the physical burden of labor has been lightened for farm workers, many small and medium-sized farms have been forced out of business.

Overall the food system worldwide accounts for approximately 28 percent of global GHG emissions. This includes preproduction activities (fertilizer, pesticides, animal feeds), production (growing crops and livestock), and postproduction activities (transport, storage, processing, retail, consumption, waste disposal). But it does not include deforestation for croplands, which contributes another 15 percent, giving a total of at least 43 percent of GHG emissions in developed countries that are attributable to food production. Some estimates put that total as high at 57 percent. Though in developing countries the level of GHG emissions is much lower than in developed countries, the portion used for food production is much higher, some as high at 80 percent.[53]

Much of food's contribution to GHG emissions involves meat and dairy products. When livestock production is considered separately, the calculations show that this sector contributes approximately 18% of global GHG emissions, bearing in mind that livestock farming overlaps in fossil fuel use with other farming activities.[54] Confined livestock production, based on feeding corn to cattle, is especially problematic. Corn-fed cattle produce unnaturally large amounts of methane, which is 20 times more effective at trapping the heat from the sun than CO_2. As the reality of climate change hits home and it becomes clearer that the use of fossil fuels must be drastically reduced, agricultural production, and the whole industrialized food system that goes with it, must change.[55]

The UN Food and Agriculture Organization (FAO) estimates that one-third of all food produced is wasted. This tremendous amount of wasted food is not only a moral issue when one billion people in the world are hungry, but it has a significant impact on climate change. The FAO estimates that wasted food accounts for 3.3 billion tons of CO_2 equivalents, or nine percent of total global carbon emissions. In developing countries most food waste happens at the production stage, while in developed countries much of the waste occurs during food processing, retail sales, and consumer end-use.[56]

Most of the GHG emissions from food waste is in the form of methane released from landfills. Although some methane is released in composting, it is still better than sending organic matter to landfills and capturing the methane from there.[57]

Climate Change and Smallholder Farmers

Of the world's 3.7 billion acres of cropland, 82 percent is not irrigated, but dependent solely on rainfall.[58] This is the case for subsistence and other smallholder farmers, many of whom farm marginal lands. The timing of rainfall is critically important for planting most annual crops. Because of climate change, seasonal rainfall patterns are being disrupted, bringing uncertainties about the timing and intensity of seasonal rainfall onset. Even without the effects of climate change, the optimal date for planting is often uncertain and planting too early or too late results in much reduced yields. With increased uncertainty over the range of optimal dates and with planting spread over a longer time period, there will likely be a loss of total production. The more unreliable the rainfall, the greater is the likely reduction of yield and output of some crops, especially grain crops.[59] Smallholder farmers may not have the capacity to withstand these shocks, especially if they are widespread, affecting most households in the region.[60]

Summary: Climate Change and Food Production

Climate change has already begun to have a major effect on food worldwide. The most devastating effects involve water: drought, flooding, and changes in the timing of rainfall, all of which are threats to food production. Increasing temperature may increase yields by increasing the length of the growing season in the higher latitudes if the crops are not already growing at the upper level of their heat tolerance.

Some food crops show increasing production with increasing CO_2 concentration, but for most crops the nutrient value is decreased unless increased nitrogen is also available. Without new heat tolerant crop strains, gains in crop yields at higher latitudes will be more than offset by losses at lower latitudes.

The overall effect of climate change is likely to be decreasing crop yields worldwide because most agriculture is in the lower latitudes where crops already grow near their heat tolerance level. Those areas are projected to experience increasing drought conditions, which will further stress food production. Food production is also seriously affected by land loss and degradation caused by a combination of climate change and human activity. This reduces the amount of land available for growing food crops and pasturing livestock. The effects of climate change, pests, and pathogens on food production are not yet fully understood, but some effects of climate change on pests are already showing deleterious effects on yields.

Food production is not the only determinant of food security, because access to food also depends on incomes, food prices, and government or community support and care programs, which are the topics of the next chapter.

Chapter Three

Control of Food Systems

No matter how much food is produced, its distribution depends on socio-political controls of land, water, and food. In this chapter we explore the questions of control: What is the role of transnational corporations? Who controls land? Who controls water? Who controls seeds? Who controls food prices? Who controls the distribution of food? What international laws, agreements and regulations have been developed to manage global resources equitably?

Control by Transnational Corporations

Control of land, water, and seeds by transnational corporations has a significant impact on the production and distribution of food throughout the world. The stage was set in the United States for the prominence of corporations in 1886 when they were given the same rights that persons are guaranteed under the fourteenth amendment to the U.S. Constitution. This precedent was reaffirmed in 1888 and many times since.[61]

The Citizens United Supreme Court decision in 2011 and the McCutcheon decision in 2014 allow corporations to engage in the funding of election campaigns without limit and without public knowledge. This means that in the U.S, corporations are dominating the electoral process, and therefore the U.S. government, especially the passage of legislation. Unlimited amounts of money are permitted for lobbying on behalf of transnational corporations. This accounts for U.S. government actions and inactions, such as, trade agreements that favor corporations, blocking progress in international climate change negotiations, and lack of federal legislative action on climate change.[62]

The industrialization of agriculture has resulted in conglomeration and cooperation among large corporations providing agricultural inputs, such as agrochemicals and seeds. Three corporations, Monsanto, DuPont, and Syngenta account for more than half the agrochemical and seed market.[63]

Who Controls Water?

Water is as vital as air in supporting life, essential in the production of food, the generation of energy, and all economic activity.[64] The ownership and exploitation of water for profit presents a fundamental conflict of interest between profit and the public good. Access to clean fresh water should be a human right. Although it has been assumed to be implied, water was not explicitly included in the original 1947 UN Declaration of Human Rights. An attempt to rectify that oversight was successful when on July 28th, 2010, the UN General Assembly voted unanimously, with 124 nations in favor and 42 abstentions, for a resolution that declared water and sanitation to be human rights. The U.S. and Canada were both among the abstaining nations.[65]

Under the European ownership and dominion concept, water rights law developed two distinct systems in the United States. In the early settlement of the eastern U.S., land taken from the Native Americans was designated as private property. There were abundant streams and rivers. A system of water law developed considering the owners of water rights to be the owners of land bordering a waterway, called "riparian" rights, which means "that lying adjacent to a stream, river, pond or small lake." The natural flow rule was that every land owner had the right for water to flow past their land undiminished in quality or quantity. Each riparian land owner had the right of "reasonable use" as long as it did not deprive the rights of those downstream. If there was an insufficient quantity available, all must reduce their use in proportion to the amount of their privately owned land. Where were the rights of indigenous peoples who first occupied that land?[66]

But in the western U.S., water was scarce and public lands were used for private purposes such as grazing and mining. A system of "prior appropriation" developed. The principles were "first in time, first in right" and "beneficial use." That meant that whoever first used the water in a particular area had the first right to it, assuming

Climate Change Manifests Through Water

Climate change primarily manifests itself through water. But it varies; different kinds of water, different ways. It can lead to more extreme weather events: either a drought or a major storm or an amount of rainfall that's unusual and leads to flooding. It's not just scarcity, it's too much, too little and unpredictably. For example, it's already difficult to predict how the monsoon season will change from year to year. It's the same with droughts in the Sahel behaving differently.

And then with salt water you have the problem of sea level rise and ocean acidification. Sea level rise is likely to devastate infrastructure along the coastlines, but it will also have a significant impact on freshwater and the economies that are tied to coastal infrastructure, which go far inland in many countries. It's not divorced from fresh water: sea level rise can penetrate aquifers that are close to the coast and then you have salt water intrusion, which makes that fresh water useless. Egypt is going to be facing a pretty significant problem in the future as a result of this, a problem that it doesn't need, as its coastal aquifers—there are many, there's a lot of water there—have sea water intrude. A significant part of Egypt's economy is based on that area of the Nile delta in and around the coastline, and a lot of the country's population is there.

—Francisco Femia interviewed by John Light[67]

that use could be deemed beneficial. Many states have evolved a permit system to enforce the prior appropriation system. Some states have systems that include both riparian and prior appropriation aspects. But these European settlers were not the first to use the water; the indigenous people were there first. And none of these systems of law considers the needs of other creatures for water.[66]

Because of the interconnectivity of groundwater systems (aquifers) and the difficulties of assessing the quantity available and removed, the issues of who has rights to groundwater is more complex. In the U.S. groundwater has generally been pumped without limits. In some legal cases groundwater has been considered as a non-renewable resource as if it were being mined. Some riparian rights have been allowed, giving absolute ownership of water underlying owned land. Some prior appropriation rights have been given to the first pumper of groundwater in an area.

On January 15, 2009, the resolution, "law of trans-boundary aquifers," was adopted by the United Nations General Assembly. Following the principles of equitable and reasonable utilization, and no harm, the resolution delineates all the factors that must be considered in the development of utilization plans, indicates that long-term benefits must be maximized, and emphasizes that an aquifer shall not be utilized in a way "that would prevent continuance of its effective functioning."[68]

Public water supply and sewage systems developed in the nineteenth century in North America and most of Europe. However, in France a private water industry developed around two major companies that have become two of the most powerful water corporations, Suez and Veolia. In 1989 Britain privatized their water systems, which led to the third largest water corporation, British-German RWE Thames. International water organizations were developed, the Global Water Partnership and the World Water Council, both of which are dominated by these major corporations.[69] As of February, 2013, "three hundred global corporations have applied to drain water from the Guarani Aquifer that sits beneath Argentina, Brazil, Paraguay and Uruguay, supplying 15 million people with water."[70]

In the 1990s the International Monetary Fund and the World Bank adopted a financial reform philosophy that required impoverished developing countries to privatize their public water and sanitation systems as a condition for renegotiating their large debt burdens. The transnational water corporations who buy these water

Water Supply in Belize

When the water supply system in Belize was sold to a transnational corporation, the services were greatly diminished almost immediately. Prices were high and the water was turned off much of the time. When it was on, the quality was low. After approximately a year of strong protests, the government started a government-owned corporation to control the supply of water, but the same transnational corporation is the majority stock holder in the government-owned corporation. Prices are still high, but water is flowing. —*Judy Lumb* supply systems expect to make a profit and the price of water has risen accordingly. Millions or even billions of women walk long distances and stand in long lines to get water for their families every day. Privatization of water has only made this worse. In Bolivia the people of Cochabamba fought against privatization of their water system and won the management of their water after two years of struggle.[71] However, privatization of water continues as the Water Yearbook reports new water companies and contracts.[72]

Strategies for eliminating corporate control at the local level are being developed. For example, the *Mora County Community Water Rights and Local Self-Government Ordinance* established a local Bill of Rights in Mora County, New Mexico. It includes the right to clean air, clean water, a healthy environment, and the rights of nature. While the impetus was to prohibit hydraulic fracturing (fracking), the ordinance also calls for an amendment to the New Mexico constitution that would elevate "community rights above corporate property rights."[73] Such civil actions are few, but important as a symbolic action even if they are overridden by state or federal laws.

Who Controls Land?

Control of land is control of food production. Private property holdings go back to Ur, one of the first city-states located in what is now Iraq. In most of Europe during the middle ages lords owned not only the land but the serfs who were attached to it. It was illegal to leave your lord's property; it made you an outlaw who could be either caught and returned or killed. The concept of private land ownership in 16th Century England was the "Divine Right of Monarchy" assuming that all land belonged to God and, as God's representative on Earth, the king had jurisdiction over all land and its inhabitants. A century later in England and other colonizing countries, constitutional government was substituted for the monarchy, and allocation of private property rights to land became the purview of governments through the issuance of land titles. As the European empires spread throughout the world, they applied this concept of titled private property to the lands that they had stolen from indigenous peoples. This concept of private property is a major factor in wealth accumulation in the developed countries.[74]

Title to property gives authority over natural ecosystems to the human or corporate owner. Environmental laws are intended to limit the amount of pollution or destruction that is allowed, but they are still part of the property-based legal system. Alternatives to property-based legal systems include: 1) common pool resource management, 2) tribal right-to-use systems, 3) leasing subject to acceptable land use, and 4) systems based on nature or ecosystems being recognized as rights-bearing entities.

For example, The Universal Declaration of the Rights of Mother Earth, presented in 2010 at a meeting held in Bolivia, acknowledges Earth is the source of life and has rights as a living being.[75] This declaration is "aspirational," that is, there is no means of enforcement.

Ecuador adopted a new constitution in 2008 that instead of treating nature as property, gave nature (*Pachamama*) legal rights, "the right to exist, persist, maintain and regenerate its vital cycles."[76] However, citing insufficient international support for the conservation plan to protect the Yasuni National Park in the Amazon, in August of 2013, Ecuadorean President Rafael Correa announced that the park would be open to oil exploration.[77]

Land Grabs

Foreign governments and investors have been purchasing large tracts of land in developing countries, commonly referred to as "land grabs," often for the production of crops destined for export rather than for local people. Governments are leasing and selling their land to corporate investors and rich nations such as Saudi Arabia, the Emirates, China, and others while displacing millions of indigenous people. Abdullah of Saudi Arabia has established a multi-billion-dollar fund to get land in other countries. Saudi Star is growing rice on 25,000 acres in Ethiopia and made a deal with the Ethiopian government for 750,000 acres more. Hassad Food of Qatar has purchased a sugar plant in Brazil and 245,000 acres in Sudan to grow rice and wheat. The United Arab Emirates are producing 350,000 tons of wheat per year on 100,000 acres in Egypt. India has acquired a total of 770,000 acres to produce rice, palm oil, and sugarcane from Tanzania, Ethiopia, Kenya, Malawi, and Mozambique. China has acquired 500,000 acres in Brazil. These conditions have led to violent regional protests in Asia, Latin America, Africa and the Caribbean.[78]

Indigenous and smallholder farmers need support as their land is being taken. Their voices must be heard if their wisdom and experience is to inform scientific advances, as we look for appropriate responses to climate change. There are over 370 million indigenous peoples living in 90 countries worldwide. These peoples have been living for thousands of years under systems of commonly held land.[79]

Concerns about the rights of Mother Earth have been directed to actions against TransCanada's Keystone Pipeline (KXL). In 2011 the "Mother Earth Accord and the KXL" was approved by leaders of communities who would be affected by the KXL. They included First Nation Chiefs of Canada, tribal government chairs and presidents, traditional treaty councils, and U.S. property owners. The Accord has now been signed by many other tribal governments, native, and non-native groups.[80]

Mayan Struggle for Land

We can see an example of the current struggles of indigenous peoples to assert their rights to commonly held land in the case of Maya traditional rights in Belize versus the right of the Government of Belize (GoB) to issue concessions to oil companies to drill on Mayan land. In 1994 the GoB declared the Sarstoon-Temash National Park on traditional indigenous lands without consulting the indigenous peoples. Despite their dismay at their land being usurped without their knowledge or consent, the indigenous Maya and Garifuna villages surrounding the park came together, established the Sarstoon-Temash Institute for Indigenous Management (SATIIM), and arranged an agreement with the GoB to manage the national park. Their management agreement allowed some use of these lands under special circumstance, but their access was otherwise limited.

In 2007 a landmark decision by the Belize Supreme Court recognizing the rights of the Mayan people to their traditional commonly held land in the Belize Supreme Court was heralded worldwide.[81] The GoB never appealed, but ignored that decision and began the process of oil exploration on Maya traditional land within the Sarstoon-Temash National Park.

–Judy Lumb

Intellectual Property

There are many existing and emerging tensions between environmental and human rights law on the one hand, and international trade law on the other. One such tension concerns ownership of, and access to, seeds. Much of this is governed by international intellectual property law, which countries are obligated to observe under the World Trade Organization rules.

An intellectual property law permits the patenting of inventions to protect inventors from others usurping and profiting from their ideas. With the development of genetic engineering, corporations have been seeking patent protection for the new life forms they have created. In some jurisdictions seeds can be patented, as in the well-known cases involving the U.S. seed giant Monsanto. Wider international law allows "intellectual property" protection for plant varieties, although it does not require countries to allow patenting of the seeds themselves. Intellectual property law is less restrictive than patent law, but still has far-reaching consequences in many developing countries as disproportionate influence continues to be given to commercial plant breeders and corporate interests.

Current seed policies favor corporate interests and do not respond to the needs of small-scale farmers and producers. Intellectual property laws especially need to recognize that the realities of climate change, critical levels of biodiversity loss, the urgent need to address food security, and increasing uncertainties about climate will lead to food insecurity and bring on violent conflict.

Both the Convention on Biological Diversity (CBD) and the International Treaty on Plant Genetic Resources for Food and Agriculture (PGR) address plant genetic resources for food and agriculture. These agreements recognize that biodiversity (including plant genetic diversity), culture, and traditional knowledge are closely linked. Ongoing negotiations within the CBD call for more recognition of local community knowledge, innovations, and practices as a vital contribution to sustainable management of biodiversity.[82]

Both the 2001 PGR and the CBD's 2010 *Nagoya Protocol* address equitable sharing of plant genetic resources and farmers' rights and recognize community knowledge around genetic resources, including for food and agriculture. The *FAO International Treaty* also recognizes traditional farmers' rights to save, use or exchange seeds. Such practices are essential to helping maintain resilient plant breeding that responds to changing climates. While it is important to acknowledge these issues, these treaties and protocols rarely have the means to require countries to observe the agreements they have signed.

Corporations try to limit these practices to create dependency on their products. Policies need to recognize that the best way to tackle food insecurity is through building vibrant local economies based on production by smallholder farmers, fishery workers, pastoralists, and other rural food workers, as well as promoting the growing trend in urban agriculture. International law and politics need to uphold and maintain local knowledge and practice, rather than work to restrict it. In particular, international trade law needs to support smallholder farming communities, their rights and dignity as they continue to feed the majority of the world's people.

Food Processing and Transportation

The transporting and processing of food have increased tremendously in the last few decades, which have added significantly to GHG emissions and climate change.[83] With increasing concentrations of people living in urban regions, more and more food must be transported over long distances. In the U.S. food travels an average of 1,500 to 2,500 miles from the farm to the table. In the U.K., food travels 50 percent farther today than twenty years ago.[84]

In addition, large quantities of food products are now shipped into North America and Europe from other regions of the planet. The expansion of this long distance global food supply system has been enabled because the full costs have not been included in the pricing. Substantial subsidies of fossil fuels have kept their cost very low and the costs of pollution have not been included. This has kept the expense of food transport and the price of food artificially low in regions of massive urbanization. This system continuously adds to the atmospheric burden of GHG emissions, and thus climate change.

Food processing has, at the same time, vastly increased in scale, which requires still greater use of energy resources resulting in even more GHG emissions. Concentrated urban populations have come to depend more and more on processed food products. Even small town and rural populations in North America have now become heavy consumers of processed foods. As processing, packaging and transporting food products became big business, the industry has increasingly consolidated into a small number of transnational corporations.[85] The following table shows the 2013 revenue and profit of the top five food production transnational corporations:

Top Five Transnational Corporations in Food Production

Transnational[a]	Revenue[b]	Profit[c]
Archer Daniels Midland	$ 89,038	$1,223
Tyson Foods	$ 33,278	$ 583
Smithfield Foods	$ 13,094	$ 361
Leucadia National	$ 9,294	$ 854
Ingredion	$ 6,532	$ 427

[a]*Data from Fortune 500*[86]
[b]*2013 Revenue in **millions** of U.S. dollars*
[c]*2013 Profit in **millions** of U.S. dollars*

U.S. Food Policy

In the U.S. there are three main policy areas that affect food from field to table:

- The Food Safety Modernization Act (FSMA),
- Free and Bilateral Trade Agreements (FTAs),
- The Farm Bill

The Food and Drug Administration (FDA) is currently in the process of making rules to implement FSMA, passed in Jan. 2011. The purpose of FSMA is to prevent or reduce risk of food being contaminated by food-borne diseases. The FDA rules are generally made for large, conventional industrial producers that gross $500,000 per year or more. Compliance with the rules is too expensive and restrictive for the smaller farms that rely on family members and neighbors' teenagers to supply the labor. Small farms that gross less than $500,000 are exempt from many of the big industrial operation requirements as long as they do not sell half or more of their product to a processor or packager who resells it. The rules for conventional, industrial operations can eat up the small farm's profits and put them out of business.

Business as Usual

In 2006 I attended the fourteenth annual session of the Commission on Sustainable Development (CSD) which was focused on energy, climate change, industry and pollution.

When the U.S. was recognized the head of our delegation said, "Thank you Mr. Chairman. I would like to introduce ... from Pfizer, who will speak for the United States." The man from Pfizer read his speech, which praised his company's employment policies and had nothing to do with energy, climate or anything else on the agenda. Then he left.

The U.S. had turned its back on participating in meaningful dialogue on problems global in scope, destructive of the lifeways of millions, and potentially catastrophic for life as we know it, and had done so in a way that shouted, "Don't expect much cooperation from the United States."

Within the CSD structure there are nine "Major Groups" considered to be stakeholders in any decisions made by the governments: Business and Industry; Farmers and Fishers; Indigenous Peoples; Local Authorities; Non-Governmental organizations (NGOs); Science and Technology; Women; Workers and Trade Unions; and Youth.

Business and Industry were alowed to give lengthy presentations and participate in formal dialogue. But there was no place for any of the other eight groups to speak a word. The response of the squelched eight was frantic collaboration to craft a statement not more than three minutes long expressing what we felt we needed to say. We were not even permitted to present this three-minute's worth of words ourselves, but we managed to get it read by a UN official so that it is in the official records.

Eight of the nine Major Groups have so much uniting us that we worked together crafting that statement. Around the world is an enormous coming together based in common purpose and cooperation. We just have to keep on keepin' on, finding what we can do, and then doing just a bit more than that. At the UN I find myself among dozens of role models that help me learn how. —*Mary Gilbert*[87]

For example, FDA rules ban free range poultry out of concern that the birds will be exposed to *Salmonella* from wild animals.[88] In reality free-range birds tend to be healthier than birds that are confined cheek to jowl in quarters that become overloaded with excrement in which bacteria are easily passed from bird to bird. The totally confined animals are fed antibiotics to prevent the spread of disease and enable faster weight gain. They also create antibiotic-resistant bacteria and, therefore, are dangerous both to the animals and to humans.[89]

The FTAs' primary beneficiaries are U.S. transnational corporations and large investors from Wall Street and their equivalents from the European Union, China, Australia, etc. Those investors and the grain, oil seed, cotton and other commodity traders—Cargill, ConAgra, Archer Daniel Midland—took advantage of federal farm subsidies under the previous Farm Bills to buy commodities at less than the cost of production. Then they could export and sell to developing countries at prices cheaper than the cost of production by local peasant farmers, putting them out of business. Investors could then buy up farms and combine with owners of land to form factory farms that sell produce to the U.S. and EU chains at less than their family farmers can produce table food, putting them out of business, too.[90]

The investor-owned factory farms have benefitted in all FTA partnerships. They use toxic chemicals and totally confined feeding methods that compromise the health of animals and pollute soil, water and air. They also collude with retail chains like Walmart, Kroger, Safeway and others to get their products favorably displayed and sold at lower than competitors' prices.[91]

FDA Rules Public Comment Session

I participated in a public comment session on April 24th. Not one of the FDA staff had ever farmed. All were academic biologists, agency administrators and former agribusiness minor executives. They have since published a revised set of rules. That set may be appropriate for large industrial factory operations, but some are not appropriate for smaller, independent, family and individually operated farms.

—Alan Connor

The 2014 Farm Bill replaces crop subsidies and price supports for crops, such as corn, soy beans, wheat, and cotton, with subsidized insurance that is available to farmers. Dairy, livestock, fruit, and vegetables also qualify for subsidized insurance. The extent to which this major change will correct some of the problems caused by federal crop subsidies is yet to be determined. Funds are allocated in the 2014 Farm Bill to help beginning farmers, to help farmers make a transition to organic farming, and to support farmers' markets. All meat labels are required to indicate where the animals were raised, slaughtered, and processed. However, funding for land conservation programs and the Supplemental Nutrition Assistance Program (food stamps) were both cut.[92]

Commodity Markets and Food Price Dynamics

Over the past forty years, local and regional traditions and customs around food have been eroded as food has become a global commodity. Currently ten percent of the cereal consumed on Earth is traded internationally. Food prices oscillate as speculators bet on food commodities.[93]

The World Bank, the International Monetary Fund, and the World Trade Organization have had a strong impact on food security. The idea was that developing countries would earn income by developing products for export. During the 1980s and 1990s, the World Bank became increasingly involved in the financing of the national debt of developing countries and they set the conditions for access to loans. To get these "structural adjustment loans," developing countries were required to:

- cut back on state-supported social, education, and health programs;
- privatize state-owned industries, public infrastructure and public utilities;
- open local markets to international investment;
- expand the export of raw materials; and
- re-orient the food sector from national self-sufficiency to export-oriented production.[94]

In most developing countries, farm credit programs were cut; seed and fertilizer supports were slashed; storage and distribution programs were eliminated. The result was that agriculture in the developing countries failed to keep up with population growth, which made them more dependent on the import of food.[95]

But at the same time many industrialized countries increased subsidies to their agricultural sectors, which enabled them to export bulk commodities at prices below the local cost of production. The flood of cheap imports made small-scale farming in developing countries financially non-viable. Farmers left the land. Many moved to cities and land ownership fell increasingly to corporate ownership, both foreign and domestic.[96]

Production of traditional food staples like millet, sorghum, cassava, and yams faltered and shifted instead to export commodities like tea, coffee, cotton, cocoa, sugar, rubber, cooking oils and timber. Africa was particularly vulnerable to this transformation as cash crop acreage organized into large land holdings for export grew to be greater than the acreage dedicated to staple foods for local consumption. Previously self-sufficient in food production, many developing countries became net importers of basic food staples and, therefore, vulnerable to increasing prices.

In 2006 staple food prices began to increase through a multiplicity of causes. Because developing countries were encouraged to reduce production for local needs, which lowered domestic grain reserves, and to rely on international trade, the in-country supply of staple food was drastically reduced. Agricultural land was increasingly dedicated to bio-fuel feed stock production, and weather disruptions—floods and droughts—decreased the supply of staples. Modern food production and transport require the use of petroleum products, and when the price of oil skyrocketed, costs increased. With that rapid change in supply and demand conditions, agricultural commodity prices also skyrocketed. The price of cereals nearly doubled within a year.

By 2008 the real estate market in the U.S. had crashed, so investors turned to food commodities in unregulated exchanges to keep their profits high. In the developing world, many families spend half their income on food; the poorest spend over three-quarters. For millions, food price escalation reversed decades of progress in reducing poverty and hunger. Many became desperate; so they took to the streets. Protests, riots, and violent conflict broke out over food price escalation across Asia, Eastern Europe, the Middle East, Africa, Latin America and the Caribbean. Demonstrations expanded to include other concerns. Commerce was disrupted. Property was

In the Midst of Food Riots

In 2008 I attended the 16th annual meeting of the Commission on Sustainable Development (CSD 16) at UN headquarters in New York, with a focus on a number of interconnected areas: Land, Drought, Desertification, Sustainable Agriculture, Rural Development, and Africa, all related to Food Security.

While we were meeting in New York there were food riots in over 30 countries. Two points of view on addressing this crisis emerged. The industrialized nations spoke for increasing food production worldwide by 50 percent or even 100 percent. They would create food security by stepping up the kind of food production used in the "Green Revolution" in India, specifically by increasing the role of agribusiness. ...

However, achieving food security doesn't necessarily include support for small farmers and nurturing a thriving local food production system. The industrialized nations would solve the problem of food insecurity by getting tons of this and that to here and there, ignoring land tenure rights, the restoration of soil to health, maintenance of forest cover, and sustainable water use.

The other viewpoint agreed that emergency food was needed right away, but providing it should be handled by agencies set up to address such crises in the short term. The business of CSD16 should be to focus on long-term solutions to the problems behind the crisis, solutions leading to debt-free economies, poverty reduction and sustainable land use.

These nations want "food sovereignty" as well as food security, where nations decide their own food production policies. They can protect themselves by refusing to let cheap food be dumped by overproducing nations. They can opt for sustainable land use and water policies, and address poverty by encouraging locally-based food production and local industry.

The serious and urgent food and climate crises are being used by political and economic elites as opportunities to entrench corporate control of world agriculture and the ecological commons." (La Via Campesina) —*Mary Gilbert*[97] destroyed. Demonstrators were killed. Governments were destabilized. Political regimes collapsed. Heads of state were murdered, prosecuted and driven into exile. Food insecurity undermined civil stability in at least 33 countries.[98]

The so-called "Arab Spring" was described as a popular movement for democratic reform. But at its base, it was a reaction to the rapid escalation in the price of food.[99]

Why did so many countries and so many people become so highly vulnerable, all at the same time, to the vagaries of weather and markets? Former U.S. President Bill Clinton provided a big clue in a speech delivered at the 2008 United Nations World Food Day, "Food is not a commodity like others. We should go back to a policy of maximum food self-sufficiency. It is crazy for us to think we can develop countries around the world without increasing their ability to feed themselves."[100]

Food Security, Food Sovereignty, and the Right to Food

The 1948 UN Declaration of Human Rights first recognized the right to food as a part of the right to "a standard of living adequate for health and well-being."[101] The concept has developed around the terms "food security" and then "food sovereignty," and come full circle with the focus on a "rights-based approach."

The United Nations Food and Agriculture Organization (FAO) defines food security as "*a situation when all people at all times have physical, social and economic access to sufficient, safe, and nutritious food that meets their dietary needs and food preferences for an active and healthy life*."[102]

The FAO derives four dimensions of food security from this definition: 1) food production, stock levels and net trade; 2) economic and physical access; 3) utilization, which means the way the body uses the nutrients in food, which covers food preparation, diet, and distribution within a household; and 4) stability of food availability over time.[103]

Each of these dimensions of food security will be impacted by climate change. In addition to the impact of climate change on food production as discussed earlier in Chapter Two, utilization of food is impacted by decreasing nutrient value of food, which results in decreasing quality of the diet. Stability in food availability and access will also be impacted as the yield in food production becomes more variable.[104]

The goal of food security could be met by dumping large amounts of food from external sources instead of from internal sources. Thus, food security is subject to manipulation by political and corporate power. The concept of "food sovereignty" was developed to take into consideration a peoples' ability to feed themselves. It was defined by the Forum for Food Sovereignty as, "*the right of peoples to healthy and culturally appropriate food produced through ecologically sound and sustainable methods, and their right to define their own food and agriculture systems.*"[105]

In 1996 the United Nations Committee on Economic, Social and Cultural Rights requested that the right to food be given more direct attention:

"The right to adequate food is realized when every man, woman and child, alone or in community with others, has physical and economic access at all times to adequate food or means for its procurement."[106]

Olivier de Schutter was appointed the Special Rapporteur on the right to food. According to him, food must be:

- Available from natural resources by growing, fishing, hunting, or gathering; or for sale in markets;
- Accessible both economically and physically, so that food must be affordable and accessible to everyone, even the most vulnerable; and
- Adequate to safely satisfy each individual's dietary needs within their cultural acceptability.[107]

The United States has been very reluctant to take up this rights-based approach to the management of food. The U.S. is one of only five countries in the world that has not accepted the right to food. Instead the U.S. continues a needs-based approach in which healthy food is a privilege of those with money to buy and any help given to those without access to food is a charitable act rather than an obligation.[108]

The U.S. Senate's reluctance to ratify international treaties is an example of the attitude of U.S. "exceptionalism" often found in political rhetoric and reflected in the Senate's inaction. However, on a practical level the Food and Nutrition Service of the U.S. Department of Agriculture administers a number of programs to provide better access to food and education programs to promote healthy eating.[109]

Food Security and International Law

It is not possible to consider our food system without looking at the structures of inequality, power and control that have been built into supply chains and that perpetuate overconsumption of resources by a small number of people in affluent regions of the world at the expense of the majority of the rest of the world. More work is urgently needed to protect the human rights, dignity and livelihoods of the world's most vulnerable and marginalized peoples throughout our global food supply chains.[110]

International environmental, human rights, and trade law are slow to respond to the future risks relating to climate change, food sovereignty and conflict. This is closely linked to the wider debate about increasing corporate control over nature, as more and more natural resources are being commodified, that is, turned into commodities to be traded in international markets. It is almost impossible to improve resilience to climate uncertainty if there is diminishing local control of natural resources as fundamental as seeds, land and water.

Since the 2007-2008 global food crisis, food security has become a top priority of social movements, including some non-governmental organizations (NGOs). United Nations agencies have hosted innumerable discussions on the best ways to respond to increasing risks of food insecurity and climate change. How do international law and policies aid the fair and equitable governance of world food systems, and do they do so in ways that respond to changing climates?

There is a range of international legal frameworks that provide standards and guidelines to support countries, communities and individuals when responding to challenges around water, land and food (*next page*). These inform and shape national laws and policies and provide an avenue for accountability and settlement of disputes.

Relevant International Laws and Standards

International Environmental Law

- *1992 Rio Declaration on Environment and Development*
- *1992 Convention on Biological Diversity (CBD)*
- *1992 UN Framework Convention on Climate Change (UNFCCC)*
- *1998 Aarhus Convention on Access to Information, Public Participation in Decision Making and Access to Justice in Environmental Matters*
- *2001 FAO International Treaty on Plant Genetic Resources for Food and Agriculture*
- *2010 Nagoya Protocol on Access and Benefit Sharing*
- *2012 CFS Voluntary Guidelines on the Responsible Governance of Tenure*

International Human Rights Law

- *1948 Universal Declaration of Human Rights*
- *1959 Universal Declaration on the Rights of the Child*
- *1966 International Covenant on Economic Social and Cultural Rights (ICESCR)*
- *1986 UN Declaration on the Right to Development*
- *1989 ILO Convention No. 169 on Indigenous and Tribal Peoples*
- *2007 UN Declaration on the Rights of Indigenous Peoples (UNDRIP)*
- *2010 UN Declaration on the Right to Water and Sanitation*
- *2011 UN Guiding Principles on Business and Human Rights*
- *2012 Declaration on the Rights of Peasants and Other People Working in Rural Areas (Open-ended Working Group)*
- *2013 Declaration of the Right to Peace (Open-ended Working Group)*

These instruments are not legally binding even for national governments that have ratified them and there is no oversight mechanisms at the international level. But they provide guidelines or minimum standards for national governments that then enact their own laws accordingly. Some of these guidelines have become widely accepted.

At the international level, food and agriculture are addressed mainly by the three Rome-based UN food agencies the Food and Agricultural Organization (FAO), World Food Programme (WFP) and International Fund for Agricultural Development (IFAD). All three agencies come together with private industry and civil society in the Committee on World Food Security (CWFS), a key committee that deals with food governance.

In recent years the CWFS has negotiated agreements and guidelines for countries to achieve sustainable agricultural development. The CWFS recognizes the central role smallholders play in securing and maintaining global food security. In May 2012 it adopted the Voluntary Guidelines on the Responsible Governance of Tenure, the first international guidelines on the issue of land tenure addressing the way legal systems govern ownership of, and access to, natural resources.[111] The guidelines are voluntary, but urge states to legally recognize customary land and natural resource rights of indigenous and local communities. The adoption of the guidelines followed three years of intensive negotiations among country governments, private industry and civil society groups including farmers, pastoralists, fisherfolk, landless rural workers and indigenous peoples. This has been reported as the most participatory process the UN has ever achieved, and many have hoped that these agreements would be used as a legal tool to prevent destructive land grabbing.[112]

International human rights law provides a rights-based approach to equitable and peaceful natural resource management. The core international human rights treaties oblige their signatory parties to respect, protect and fulfill the human rights to water and food, as well as access to information, public participation and justice. The right to food recognizes the fundamental right of everyone to be free from hunger. The human right to water acknowledges "sufficient, safe, acceptable and physically accessible and affordable" water as essential to leading a life in human dignity.[113]

The various UN human rights bodies have reporting and monitoring procedures, as well as special rapporteurs on each of these rights to help countries live up to their aspirations in these agreed documents. Olivier De Schutter, who was the UN Special Rapporteur on the Right to Food at the time of drafting this pamphlet, stated that the right to food is inseparable from access to productive resources. Security of land tenure and control over plant genetic resources is essential for long-term food security.[113] A 2013 study by the Human Rights Council Advisory Committee identifies small-scale farmers, landless farm workers, fisher-folk, hunters and gatherers as among the most discriminated against and vulnerable people in many parts of the world with thousands of peasant farmers becoming victims of forced evictions from land every year.[115]

The UN Declaration on the Rights of Indigenous Peoples (UNDRIP) was adopted in 2007 by the UN General Assembly and was the first universally accepted declaration of indigenous peoples' rights. It calls on States to consult and cooperate with Indigenous peoples to obtain their ***Free, Prior and Informed Consent (FPIC)*** before adopting and implementing measures that may affect them, particularly with respect to projects involving the development, use or exploitation of natural resources.[116]

The UN Principles on Business and Human Rights were developed by the UN Special Representative on Business and Human Rights and adopted by the UN Human Rights Council in May 2011, after five years of negotiation. These guiding principles call for states (nations) to protect against human rights abuse by third parties, including business enterprises. The obligations of businesses to respect human rights and address adverse impacts on human rights as a consequence of their business activities are now publicly defined and internationally agreed. Unfortunately, there is no enforcement of these guidelines except at the national level and transnational corporations may not be governable by national laws.[117]

Although few of the international agreements discussed above are legally binding in a way that is easy to enforce, they provide useful frameworks from which to consider prevention of conflict around natural resources. Much work is needed to bring these principles to other fora within and outside the UN. Several of the instruments offer opportunities for countries to share information, resources and recommendations, and provide opportunities to receive capacity-building and technical assistance from UN agencies. Many guidelines are useful for raising awareness at local and national levels and can be part of the empowerment process that leads to local groups being effective partners in natural resource management that responds to intertwined issues of climate change, food security, and the risk of violence.

Summary: Corporations Control Resources

While there may be enough food to feed the human population in the next forty years, and even that is uncertain, what is lacking is basic human fairness, appropriate controls on the market system and corporations, and management of resources with the participation of all stakeholders. International trade laws have the effect of keeping control in the hands of transnational corporations. The effect of these laws on seed policies is especially deleterious for small-scale farmers who produce most of the global food supply. Mechanisms to counter this and establish more equitable controls include declarations such as Rights of Nature, Right to Food, and the Rights of Indigenous Peoples. However, these declarations are legally binding only on those countries that have signed them and the U.S. has failed to sign most of them.

The next chapter investigates the likelihood that without adequate laws and agreements these situations of control, conflict, and marginalization will lead to violent conflict.

Chapter Four

Violent Conflict: The Impact of Climate Change

Differing views, ideas, interests, needs and expectations of outcomes that inevitably arise between individuals, groups, communities and nation states may lead to conflict. When essential resources that sustain life, such as, food and water, are threatened, that conflict can become destructive and violent. Although there are many causes for conflicts that impact human security, climate change and food security are important factors that threaten cultural identity, ecosystems, livelihoods, migration, health, and safety.

Some existing violent conflicts presently occurring in Syria, Iraq, the Central African Republic and South Sudan have resulted in the most severe migration of humans since data has been compiled. During 2013, violent conflict has resulted in 10 million additional refugees and internally displaced persons who have left their homes and communities, which was the highest annual displacement since the UN High Commission on Refugees began keeping records in the 1950s. This brings the global total of displaced persons to 50 million at the end of 2013.[118] Although, this migration cannot always be linked to climate change, refugees are directly impacted by climate change effects on the availability of food. Migration is exacerbating existing conflicts that undermine political and social stability. Add to this situation the scarcity of food and water, and the failure of global governance systems worldwide to resolve differences, and violent conflict will surely continue to escalate.

Populations that are unable to migrate face higher exposure to weather-related extremes relating to food and water scarcities. "People who are socially, economically, culturally, politically, institutionally, or otherwise marginalized are especially vulnerable to climate change and also to some adaptation and mitigation responses."[119]

This chapter investigates the potential for climate disruptions to cause new conflicts, or exacerbate already existing ones, and looks at mechanisms to prevent conflict from becoming violent.

Food, Water and Violent Conflict

Hunger and conflict over water resources leading to violence is nothing new. But today that global threat has an entirely new magnitude, scope, and persistence. The impact of climate change over the past decade has already created zones of conflict and reactions of violence. Even worse impacts projected by scientists for the near future will produce resource conflicts in breadth and scope beyond any known in history. Climate change is likely to exacerbate existing problems and "push disputes from peaceful negotiations to violent confrontations."[120]

The U.S. Department of Defense is taking climate change very seriously and has set out a list of recommendations on how to cope with the impacts of more severe weather, drought, and sea level rise. The Defense Science Board states in their 2011 recommendations to the Secretary of Defense:

Changes in climate patterns and their impact on the physical environment can create profound effects on populations in parts of the world and present new challenges to global security and stability. Failure to anticipate and mitigate these changes increases the threat of more failed states with the instabilities and potential for conflict inherent in such failures.[121]

The Defense Department study identifies developing countries as being the most vulnerable to the detrimental effects of climate change, as these are places where infrastructure for dealing with any kind of resource scarcity is minimal and people are already living on the edge or beyond. Further, the study highlights fresh water as the resource most likely to be impacted by climate change with water conflicts leading to warfare and human suffering.

The 2004-2009 genocide in Darfur was cited as the first occurrence of large-scale violence resulting from the effects of climate change.[122] Prior to drought conditions, nomads and farmers coexisted, sharing resources. Since 1984, drought conditions have prevailed, with the growing season shrinking from five months to three months. Due to the drought, indigenous African farmers restricted water to Arab nomads, which led to ethnic violence. Resource refugees fled to Chad, resulting in further violence in Chad and more resource conflicts. This recent history of Darfur is a cautionary tale for many more regions as climate change impacts intensify.[123]

Evaluating both the severity of the impact of climate change and the level of already existing tension can identify areas that are expected to experience the greatest threat of violence. Forty-six countries have been identified where the effects of climate change interacting with current social, economic, and political problems create a high risk of violent conflict. An additional 56 more countries are struggling with the effects of climate change. These countries have a high risk of political instability and violent conflict in the future.[124] Some examples of the worst situations are described below.

With a 3.6°F increase in global temperature, and the resulting sea level rise of one meter, ***Bangladesh*** is expected to lose 17 percent of its land.[125] This will displace 20 million people. If climate change results in global temperature rising more than 2°C, the number of displaced persons from Bangladesh is likely to be closer to 200 million. Wherever they go, this extent of mass migration is huge. Most of these displaced persons will head for India, a situation not lost on the Indian government. They are already building a border fence. "The potential for violence along the border is real. The combination of desperate refugees on one side and a country trying to prevent massive illegal immigration on the other is a recipe for conflict."[126]

The ***African continent*** will be hit the hardest by climate change. Poverty is already increasing; a large percentage of the economy relies on agriculture; and droughts are becoming more severe resulting in desertification in the north. Unprecedented rainfall is causing disastrous flooding in the south. In ***Kenya*** deadly conflicts between farming and cattle-herding people have erupted, killing hundreds. ***Nigeria*** in West Africa is particularly vulnerable to violent responses to the effects of climate change because of already existing ethnic tensions and a weak and fractured government.[127] The Nile River provides water to over 300 million people in ten countries in ***East Africa***, all of which are vulnerable to deleterious effects of climate change on their water supply.[128]

Water resources are diminishing in the ***Middle East***, which is already suffering from limited water supplies. Two particular sources of tension are: 1) Turkey controls the headwaters of the Tigris and Euphrates Rivers, which supply water to downstream Syria and Iraq; and 2) the Jordan River is the major source of water for Jordan, Israel, Syria, the West Bank and Lebanon.[129]

In ***Palestine***, the Israeli Army has destroyed water tanks and wells in the occupied territories. In ***Indonesia*** there was deadly violence over a water source in Maluku.[130]

With the melting of Himalayan glaciers, ***Pakistan***'s main river valley, the Indus, is expected to first experience flooding, and then water shortages. The Indus flows out of India into Pakistan. During political crises between the two states, India has regularly used its control of water resources to punish Pakistan. Concern has been raised that "if India does not abstain from its course of action, there would be a nuclear war over the water dispute."[131] In addition, several other major rivers flowing from the Himalayan glaciers into India, Southeast Asia and China will be drastically diminished in volume, which will increase conflict over water resources in the most densely populated regions on Earth.

Syria suffered a five-and-a-half-year drought from 2006 to 2011 that had unprecedented devastating effects. Nearly 75 percent of the farmers in the northeast suffered total crop failure. Herders in the northeast lost around 85 percent of their livestock, which affected about 1.3 million people. Herders and farmers in the north and south had to pick up and move. Millions of farmers and herders were trekking into urban areas. Iraqi and Palestinian refugees have been flowing into Syria since 2003. Syrian cities were already hard-pressed economically. High food prices triggered deadly violence, which led to the migration of refugees into Lebanon, Jordan, Iraq and Turkey. While climate change may not have caused these conflicts, it is what the security community calls a "threat multiplier," that makes other threats to human security worse.[132]

In ***South America***, struggles between governments, wealthy families, transnational corporations, indigenous and other poor people over water being used for industrial purposes such as mining are turning deadly. According to the FAO, this region has more water than any other region on earth, with 29 percent of the world's reserves. However, the use of water by industries, such as, mining and other commercial uses, impacts the survival of families and their ability to grow food and have clean drinking water. This is fueling violent protest. International corporations have been known to hire mercenary armies to enforce their claim to traditional lands occupied by indigenous and other poor people, usually with complicity of the specific national government.[133]

Mining operations use tremendous amounts of water and rainfall has decreased drastically, especially in the mineral-rich highland areas. Deadly violence erupted in ***Peru*** in 2010 during continuing protests over diversion of water for new gold and copper mines.[134] Similar issues are occurring between indigenous tribes and the government of ***Ecuador***. The tribes are preparing for war in protest over the government's approval of the drilling for oil on eight million acres in the Amazon Rainforest. The tribes' expulsion from their cultural homeland and way of life, and their possible extinction, are due to structural violence perpetrated by government against citizens.

These are only a few of many such examples that might be cited where water and food, while not the sole cause of conflict, are significant contributing or exacerbating factors. Between them, governments and transnational corporations with shared economic interests are overriding the human rights to food, water and security worldwide. According to Food and Water Watch, "around the world, multinational corporations are seizing control of public water resources and prioritizing profits for their stockholders and executives over the needs of the communities they serve."[135]

But there are examples where the people have taken hold of their own destiny. In Cochabama, ***Bolivia***, massive protests in 2000 were successful in reversing the privatization that was required by the World Bank. Privatization of their water resulted in tripling or quadrupling the water bills, requiring as much as half the monthly income for some. Protests shut down the city for four days and protesters suffered civil rights abuse when martial law was declared, but in the end the government terminated the contract and gave control of the water system to the protester's organization, Coordinadora de Defensa del Agua y la Vida.[136]

Jointly managing water resources can also be a pathway to peace. The management of the Mekong River between ***Vietnam***, ***Cambodia*** and ***Laos*** during the Vietnam War provides another encouraging example.[137]

According to the drought information system which monitors global conditions, drought continues to be intensified and expanding in North and Central America. In the ***United States***, over fifty percent of the country is experiencing different degrees of drought and it is expanding. These conditions globally will have a profound impact on food production, pricing and populations. Vulnerable populations in the developing world, as well as the rural and urban poor are most at risk.

Global Governance and Violent Conflict

Conflicting interests and how to deal with them in a non-violent way is the focus of this section. When conflict arises among nation states, the complexity of geopolitical issues, cultural differences, historical relationships, resource access and economic interest provokes protective attitudes that can harden responses so that conflict becomes violent. In a clash over who controls water and land, those disaffected may turn to weapons and violence aimed at their perceived oppressors, be it transnational corporations, governments, or neighbors. When this happens, and often before situations reach this point, those in power may resort to violent means to suppress others whose growing strength and anger they fear, resulting in deadly confrontation. Existence depends on access to essential resources such as food and water, so deprived humans will resort to violent means in order to survive. The powerful will do likewise in order to maintain power over popular initiatives that they perceive as threatening their dominance.

Universal declarations, treaties, and resolutions under the jurisdiction of the United Nations serve as models with established guidelines and some international law regarding human rights. But the primary responsibility for enforcing these laws and guidelines rests with the nation states, which must have the national legislation to take legal action and provide agencies for enforcement.

Descent of a Once Prosperous Land

"The Albertine Rift is that part of Eastern Africa where plate tectonics are pulling the land apart. With rich volcanic soils, abundant rain and numerous lakes, rivers, mountains and prairies, this land supports an extraordinary flourishing of biodiversity. The land supported both farmers and pastoralists, but well-organized monarchies spread throughout the region, establishing the firm dominance of the minority cattle herders over the agriculturalists. Impressed by their cultural sophistication, Europeans settlers of the late 1800's not only exploited this class division, but also froze it into place by issuing identity cards. The fifteen percent who owned cattle were defined as Tutsi and the farmers were defined as Hutu.

"National independence in the 1960's created several new nation states along with heightened competition for political influence and privilege throughout the region. A declining death rate and traditionally high birth rates caused a demographic explosion that had quadrupled the population over the past fifty years. Human population densities soared. Division upon inheritance reduced the average plot size to an insufficient half acre. Over-reliance upon export crops with collapsing prices undercut household economies and the ability to steward the land. Overworked and overgrazed, the land became increasingly infertile.

"With both hunger and impoverishment, politics fell readily into the hands of extremists. The Hutu revolted against their Rwandan overlords. Tutsi reprisals and subsequent actions and reactions precipitated both civil war and mass genocide. In just three months, 800,000 were dead and 2,000,000 Rwandans, mostly Hutu, fled the country.

"Western observers attribute the atrocities in part to the fact that there were too many people on too little land. It is commonly heard among Rwandans today that the war was necessary to get rid of too many people, to bring numbers down to what the land could support." —Robert Draper[138]

Should we accept such easy explanations? While it is easy to interpret the Rwandan genocide in Malthusian terms, a more nuanced explanation would include distortion of right relationships among the people and toward the land. —Phil Emmi

Attempts to pre-empt violent, potentially deadly, conflict require the identification of vulnerable groups by governments and commitment to respond to their needs. Class or caste differences at the local level might argue for intervention by governments. Conflict resolution between nations, such as, disputes over water or land rights, is the purview of the United Nations General Assembly and Security Council, as well as regional organizations, such as, the African Union and the Organization of American States, although to date they have had very limited success.

In 2012 there was a total of 396 conflicts, 43 were classified as highly violent and included 18 wars and 25 limited wars; 208 involved some form of violence.[139] In 2013 that total had rise to 414 global political conflicts, 221 of which had been violent in earlier years; there were 20 wars in five regions of the world, 25 limited wars, and 45 highly violent situations. There were 176 violent crises and 193 non-violent conflicts.[140] Often, the underlying causes of conflict involve issues of food, water or land.

Both food insecurity and deadly conflict inevitably result in refugees. The number of conflict-induced displaced persons reached 50 million worldwide by the end of 2013 according to the UN High Commissioner of Refugees. Most (86%) were hosted by developing countries.[141]

Displaced persons, whether internal or external, have fled the land where they grew their own food or had the ability to buy it. Providing adequate food for refugees is a major problem in the world today. Refugees displaced by violence are almost always impoverished and food insecure. In this situation female children are sometimes married off to older men because the family cannot afford to feed them—another source of violence resulting from food insecurity.

Food policies, agreements, regulations, use of productive land for alternative uses and markets are some factors sensitive to climate variability and climate change. Political stability is vitally necessary for governments to meet the needs of their populations. Corruption and favorable treatment in some countries directly affects the food security of the most vulnerable. In some parts of the world, including the Middle East, North Africa, the Andes and the Caribbean, skewed water policy allocation may favor the affluent, which heightens the vulnerability to climate stress. Certain adaptation actions implemented by governments in the areas of property rights and conflict management can lead to more conflict. The efforts of inefficient or even illegitimate institutions to mitigate or adapt to climate change can change the distribution of access to resources and have the potential to create and aggravate conflict.[142]

An Unspoken Form of Violence

Regardless of how it comes about, the experience of food insecurity constitutes a form of oppression. Be the oppressor a vengeful god, a feudal lord, or an unnamed market force, the oppressed internalize an image of the oppressor, what the oppressor wants, and the rules that must be obeyed. For the oppressed, the capacity to see oneself as having value, to claim a right to basic human dignity, to honor one's own judgment, to exercise a sense of independence, to act on the authority that comes from within, and to contribute to society according to one's own unique gifts lies beyond the realm of possibility. For the oppressed, being free and helping others free themselves dare not be simply an unfulfilled aspiration: it is rather a condition to be feared. Freedom is not something simply given by oppressor to the oppressed. It needs to be won in the course of further human values development.

No matter how mature and peaceful they might otherwise be when not so stressed, people under the stress of food insecurity, especially when feeling discounted and undervalued, are likely to revert to behaviors for survival that can threaten violence. And those seen as discounting them, unwilling to support them by sharing, will likely respond in fear with pre-emptive violence. The history of hunger accompanied by violence is a long history.

Summary: Climate Change is Exacerbating the Threat of Violent Conflict

While much hunger now present in the world is directly caused by human action, climate change will increase the number of those experiencing food insecurity and hunger, with potential for massive migrations further aggravating social instability and the potential for violent conflict. Conflicts over water are turning deadly today and are likely to be exacerbated by climate change.

Preventive measures are possible but they require effective government capacity with monitoring and preparedness to respond to the plight of vulnerable communities and international support for governments under stress. International and national frameworks and capacities need strengthening to become effective in ameliorating hunger and countering potential for violence. Peaceful resolution can be a choice if sufficient political will can be generated.

Measures needed to adapt to the impact of climate change with the goal of preventing violent conflict are discussed in the next chapter.

Chapter Five

Responses to Food Insecurity from Climate Change

Can humankind adapt quickly enough in the next four decades to minimize the effects of climate change, feed the increasing global human population, and avoid violent conflict? In this chapter we explore mechanisms for addressing climate change, for adapting agriculture to new situations brought by the changing climate, caring for vulnerable populations, and for managing natural resources in a fair and equitable manner. The goal is to create a world of shared resources and security that will reduce the likelihood of violent conflict.

Addressing Climate Change

The 2013-14 IPCC fifth assessment makes a strong case for the urgency of acting to reduce GHG emissions. Summarizing the results of all the recent economic analyses of future scenarios, the assessment concludes that the longer the inevitable changes are put off, the more it will cost, both for making changes and managing damages. Estimated costs by the end of this century will be four to ten times higher without serious reduction of GHG emissions by 2030. "The report emphasized that the world's food supply is at considerable risk—a threat that could have serious consequences for the poorest nations."[143]

"Stabilizing GHG concentrations will require large-scale transformations in human societies, from the way that we produce and consume energy to how we use the land surface." Chapter 6 of the Working Group III report explores ***transformation pathways*** towards stabilization of GHG emissions. These pathways include "accelerated electrification of energy end use, coupled with decarbonization of the majority of electricity generation and an associated phase-out of freely emitting coal generation." This means switching to electric vehicles as quickly as possible to eliminate the use of liquid fossil fuels, such as gasoline and Diesel, including bio-diesel and switching to renewable energy for generating electricity to eliminate coal power generation. Renewable energy costs have gone way down since the previous IPCC assessment in 2007, which makes some of these transformations much more economically feasible.[144]

Life style changes, especially in the U.S., are urgently needed. Eating less meat, buying locally, and decreasing consumption by repairing, recycling, and reusing will all help to conserve energy and natural resources, and reduce GHG emissions. A cultural change from a consumer society to one of community cohesion will also address inequality issues and increase general happiness.

Land use issues are often overlooked, but constitute a large part of the GHG emissions equation. The absorption of GHG by trees is a major factor, so deforestation greatly increases GHG accumulation. Deforestation must be reduced or eliminated altogether and afforestation increased. Plant trees!

Adaptation to Climate Change

Industrialized nations, having failed to account for the full costs in their accumulation of great wealth, have destabilized the earth's climate and accrued a large "ecological debt" to developing countries. Developing countries have already suffered and will continue to suffer much more from the results of climate change than the industrialized countries. The countries suffering most have the fewest resources to adapt to their changing situation. For both moral accounting and shared security reasons, the industrialized countries must now ***provide funding and appropriate technology to developing countries*** for adaptation to the effects of climate change and compensate them for their losses and damages. Among other mechanisms is the Green Climate Fund, which developed countries must support.

Adapting Agriculture to Climate Change

Adaptation to climate change will mean ***changes in which crops and livestock*** are grown in particular areas, and ***changes in management*** of those crops and livestock, as well as development and restoration of the use of plant varieties that have the needed climate-adaptive characteristics. In adapting to new conditions, farmers must distinguish between ordinary weather variability and fundamental changes in the climate—changes that are unlikely to settle into reliable weather patterns as global warming continues. They might need to plant at different times or switch to different crops. Changing any of their methods or crops involves risk, so insurance programs should be made available to support these adaptation measures.

As temperatures increase, the ***growing season length*** increases in higher latitudes, so some areas that previously were not conducive to raising particular crops like corn and wheat may become favorable. But those same areas will very likely not have the appropriate soil as ice age glaciers pushed the soil built up over millennia away from the poles, thus amassing the deep, rich topsoil of the prairies in North America and similar fertile areas of Eurasia, and leaving areas to the north stripped of good soil. As food production moves north, particular attention must be paid to soil building and soil enriching practices in order to ensure continued productivity.

As they adapt to climate change, farmers will need to change their method of ***irrigation***, or develop systems to begin irrigating. Improvement or new irrigation infrastructure has historically required government financing, usually through contracting the work out to international corporations primarily interested in profit. However, even if public financing is available, the question is the availability of sources of water to be used for irrigation. Ground water is being depleted as water is removed faster than it is replaced by natural means, and the increased use of fracking for natural gas threatens to pollute groundwater sources.

Over the next thirty years, the availability of clean water for irrigation will become increasingly critical. Management of water as a common resource could maximize its use for the common good. Examples of long-term community management of irrigation systems include traditional indigenous systems in the Andes,[145] a system in Spain in continuous operation since the year 1435, and one in the Philippines since 1630.[146] The longevity of these systems of management by local self-governing institutions is attributed to a set of rules that are described later in this chapter (*p.65*).[147]

Indigenous people and smallholder farmers must be included in the design and execution of all agricultural development projects. They can provide very important information and are more likely to implement the results of the research when they are part of the design team. Current practice is for research to be done by academic institutions and biotechnology companies with very different goals and perspectives from those who are feeding most of the human population. Often, scientific research has ten-year horizons for testing hypotheses. Farmers can learn from year to year with greater adaptivity than that which science provides.

There is a global trend toward ***corporate ownership of seeds*** and plant genetic resources. This raises serious concern over availability of seeds and pricing. Patenting and intellectual property law are restricting local seed networks and having a negative impact on the diversity of crop varieties available, the maintenance of local knowledge systems, and the continuing viability of associated farm-based cultures.[148]

A more highly adaptive response to climate uncertainty is demonstrated in ***resilient local seed networks*** found across the world. How significant are these informal seed systems in contributing to food sovereignty? Half-a-billion small-scale farms averaging less than five acres provide the majority of the world's food. Farmer-developed crops have contributed at least 1.9 million varieties to seed banks worldwide.[149]

Local seed networks allow flexible and adaptable plant breeding that is continuously evolving through the saving and exchanging of seeds from one harvest to the next. In this way plant varieties can respond to changing local conditions and climate uncertainties. For example, drought-resistant crop varieties can be developed from old seed stocks that already require less water. This search for optimum diversity is maintained by small farmers every season.[150]

Local seed systems and the knowledge associated with them are central to the livelihood viability of small-scale farming cultures, and to securing a stable source of food and nutrition. Unrestricted, community-based plant breeding ultimately maintains a genetic base vital for the future food sovereignty of everyone on the planet.[151]

Sustainable agriculture must be developed worldwide and extended indefinitely in order to insure food security for all people and preserve the bio-productivity of Earth's ecosystems.

Agro-ecology is a whole-systems approach to agriculture and food inspired by ecological systems and incorporating traditional knowledge, farmer's experience, organic agriculture, and local food experiences. In developing countries traditional farmers have inherited systems that were developed over centuries to produce food under their local conditions without modern equipment, chemicals, or bio-engineered seeds. These systems often exhibit high biodiversity, ingenious technologies for management of land and water resources, diversified agricultural systems that exhibit resilience in the face of change, strong cultural values, and collective social organization. Agro-ecologists strive to learn from these systems and to reverse the unsustainable trends in modern agriculture.[152]

The objectives are to link ecology, culture, economics and society for sustainable food production and healthy environment, supporting vibrant food and farming communities worldwide. Farmers practicing agro-ecology use renewable sources of energy, biological nitrogen fixation, naturally occurring materials, and on-farm resources. They recycle on-farm nutrients, conserve soil by using organic matter for fertilization, minimize soil erosion by using perennials and mulch, and practice soil management methods that reduce plowing and tilling. They conserve water by depending only on rainfall or using efficient irrigation methods, and conserve agricultural genetic diversity by saving seeds, exchanging seeds, and using heirloom varieties. They manage the landscape by maintaining undisturbed areas and waterway banks as buffer zones, using intercropping, cover cropping, contour tillage, no-till, and rotational grazing.[153] Most national governments work with the UN programmes and have national agencies to spread use of the best of these techniques to farmers not familiar with them.

Deflate the Power of Corporations

Trade agreements have greatly increased the power of corporations in recent years. Citizen engagement with the political process around U.S. Senate consideration of trade agreements is urgently needed to prevent them from being fast-tracked through to approval without civil society input. Safeguards protecting the public interest

An Ecological Dairy Farm

Francis and Susan Thicke own and operate an 80-cow, grass-based, organic dairy near Fairfield, Iowa. They have a processing plant on their farm where they produce bottled milk, cheese and yogurt which they market through grocery stores and restaurants in their local community. All their milk products are sold within four miles of their farm.

A grass-based dairy farm illustrates one way to mimic the prairie ecosystem. In a grass-based dairy, the landscape surrounding the milking barn is converted into a polyculture of grasses, legumes and forbs—some of which are planted and some that "volunteer." This landscape of perennial plants is divided into paddocks using inexpensive fencing materials, with cow lanes connecting all paddocks to the milking barn.

After each milking (twice a day) the cows are allowed to graze a new paddock area that is just large enough to provide the cows' forage needs until the next milking time. As the cows rotate through the paddocks, grazed areas have time to recover, allowing plants to regrow to a stage of optimum nutrition for the next grazing episode.

Management is important. If paddocks are allowed too much recovery time, the plants will become overly mature and lose nutritional value. If grazed again too soon, some plant species will not recover fully and die, reducing pasture productivity and diversity. Under good management, plant diversity is maintained or increased and soil fertility is continuously regenerated.

When cows are kept in confinement, the cows' forage must be mechanically harvested in the field, hauled to the facility, stored, then taken out of storage each day to feed the animals. The cows' manure must be collected, stored and eventually hauled back to the fields. All these operations require fossil fuel energy.

By contrast, a well-designed grass-based dairy accomplishes the same objectives by the farmer simply opening the gate to the next paddock. The cows harvest their own forage and spread their manure. And they enjoy their work!

—Francis Thicke[154]

must be included. As this pamphlet goes to press, two such agreements are being negotiated, the ***Trans-Pacific Pact Partnership (TPP)*** and a similar agreement between the U.S. and the European Union (***Transatlantic Trade and Investment Partnership, TTIP***). Concern has been expressed at the secrecy around these agreements and leaked indications of benefits only to corporations.[155]

The increasing concentration of power in corporations is a direct threat to democracy. The ***Citizens United and McCutcheon decisions*** of the U.S. Supreme Court have removed the limits to the amount of money corporations can use to control the U.S. Congress, and the requirement to report sources of campaign funds. A constitutional amendment to reverse these decisions has been proposed and will require widespread political support to be passed by Congress.[156] Another political effort is ongoing to overturn the precedent that has given corporations the same legal status as persons.[157]

A movement has been growing to reform the legislation that governs the granting of corporate charters and allow ***Benefit Corporations*** in an effort to require greater accountability to the public interest. The problem is that current corporate charters stipulate that the corporations' first obligation is to their shareholders, so if the corporations take actions for the common good that sacrifice some of their profit, shareholders can sue them. The new Benefit Corporation removes that stipulation and requires corporate action for general benefit. As this pamphlet goes to press, 25 U.S. states have passed laws establishing Benefit Corporations that are "required to create a material positive impact on society and the environment and to meet higher standards of accountability and transparency."[158]

Forgiving International Debts

Many developing countries are indebted to wealthy countries and financial institutions with such high levels of debt with high interest rates that they cannot provide the services needed by their own people, including those related to food production, agriculture, and food security. Conditions imposed to refinance these debts often impact food security, such as the requirement to privatize the water supply. Spearheaded by Jubilee USA, new calls for forgiveness of developing country debts have come from many faith-based groups and other civil society organizations.[159]

Strengthen Social Safety Nets

In order to ensure food security in times of crisis or severe adaptation stress, societal safety nets need to be strengthened. The safety net created in the United States in the 1930s worked well to decrease inequality, but has largely been dismantled. Further dismantling is threatened in the U.S. Congress as this pamphlet goes to press. Similarly, the U.S. "War on Poverty" in the 1960s and programs set up to address hunger and food insecurity were largely successful until the 1980s when the Reagan administration began massive roll-backs of public support.

Existing U.S. programs such as SNAP (food stamps), Social Security, and Medicare must be supported to avoid funding decreases or changes that interfere with their purpose as social safety nets.

In regions of rural poverty, access to off-farm income can help buffer the negative effects of climate change on food production. The urban poor, even in wealthy regions, often spend as much as 60 percent of their income on food. Increasing food prices, resulting from climate change, will become a prolonged crisis for much of the world's population, both rural and urban.

Systematic, proactive responses to food security issues and crisis scenarios should be developed at various levels of society. Governments must take the lead by coordinating societal resources for adaptive responses. Serious consideration should be given to the creation of "operation rooms," which are effective tools for implementing government action.

The ***operations room*** concept derives from the physical creation of a room with wall charts plotting the progress of indicators. Operations rooms are staffed by members of ministries/departments that have roles in responding to signals demanding action. Operations rooms address the need for governments to be able to respond speedily to emergencies and to chronic, but changing, conditions where several ministries are likely to be involved in taking action. Officials from the relevant ministries are involved in designing the system. A key task is the identification of vulnerable groups and their specific circumstances. Indicators of status, trends, and the forces or events driving them are identified with threshold values that indicate the need for action. Responsibilities are assigned to monitor and respond to specific indicators of crisis. For example, food prices might be monitored and trends in loss of livelihoods or income adequacy among specific sectors of the population assessed. Indicators would be plotted on charts updated and displayed in the operations rooms. Specific responses to the threshold values are pre-planned with training, material, and budgetary provision. A signal-response system is set up that would trigger indicated action by several ministries under a coordinated budget and oversight authority. Government and societal response set up in this way can plan and implement required interventions for food security breakdowns.[160]

Empower Local Communities to Manage their Resources

Governments should play a facilitative role in managing environmental resources, valuing and promoting approaches that reduce and equalize power imbalances and lead to consensus decision making. Equitable and sustainable management of land, water, seeds, forests and fisheries are vital to prevent violent conflict over natural resources. This includes equitable access to these resources for the billions of people whose livelihoods, identities, and wellbeing depend on it. The profound power differences between corporations and local communities in cases concerning natural resources must be addressed.

Overt violence is the epitome of conflicted relationships, but there are also structural relationships within and between societies that, while not overtly violent, are clearly violent in their effects. Violence can also be embedded in the structures of society through pervasive social injustice, chronic inequality, enforced poverty, and power imbalances. Corporate control and expropriation of local natural resources that cause damage to local communities' access to clean water, adequate food, and health are examples of structural violence.

For example, hundreds of thousands of acres of land are being leased for agricultural plantations by both agri-business corporations and foreign governments, displacing local communities from the land. With this displacement, local control over food is diminished as smallholder farmers are forced into poverty, often leading to unwilling migrations to cities. Access to critical resources like water and seeds is made increasingly difficult due to "commodification," the creation of commodity markets of essential resources.[161]

In addition to the conflicts engendered by their encroachment on the lands and cultural integrity of human communities, and the structural violence imposed by the destruction and impoverishment of local food economies, a new kind of war—resource war—is explicit in the "new economic order" of global capitalism and its state-assisted hegemony. This changing reality adds a new field of action to the work for justice, peace, and human rights—the field of "food systems."

The globalized food economy has accelerated consumption in the global North and depletion of resources in the global South. The competition for the control of land and depletion of resources is leading to increased tension and conflict within local communities.

There is a clear link between the consumption-driven North and the depletion of the South's resources. Many foreign investments in the South are driven by consumption in the North. Climate change now enters the network of relationships that binds North and South into a conflicted community. Climate change has been predominantly driven by northern industrialized societies, while the deleterious effects are falling first on the economically struggling regions of the south.

Participation of All Stakeholders in Governance of Resources

Since climate change has already begun to cause resource scarcity, future effective governance of common resources at national, local and community levels is crucial. The success of governance depends on the extent of cooperation between these three levels of government and allocation of management authority to appropriate institutions.[162] Poor governance might involve lack of capacity, inadequate policy goals, and constraining power realities of conflict of interest. Capacity includes effective processes for decision-making, implementation, management, and accountability, engaging the right people in appropriate roles with relevant competencies and appropriate financial/material support.

Effective governance of common resources depends on the involvement of stakeholders at all levels in making decisions and establishing policies. Community involvement is especially important. Power dynamics within communities and with regional and national institutions must be understood to ensure community involvement. A study of citizen participation defined gradations of community involvement. The lowest level was "non-participation" in which citizens experience only manipulation. The next level was "tokenism," which includes information, consultation, and placation. The highest level was "true citizen power," which involves partnership, delegated power and citizen control.[163] One measure for involvement is to have the administration secure feedback on its concerns, objectives, and intervention proposals. Approaches to peace-building among competing interests include local leadership, creation of dialogue between groups, and empowerment of vulnerable groups.[164]

Throughout the world, methodologies have been developed to facilitate the participation of stakeholders in natural resources management. A few examples are given below:

Free, prior, and informed consent (FPIC) is the principle that a community has the right to give or withhold its consent to proposed projects that may affect the lands they customarily own, occupy or otherwise use. FPIC requires informed, non-coercive negotiations between investors, companies and governments with indigenous peoples prior to the development and establishment of oil palm estates, timber plantations, big dams, mining, oil extraction, or other enterprises on their customary lands. This principle means that those who wish to use the customary lands belonging to indigenous communities must enter into negotiations with them. It is the communities who have the right to decide, using their customary systems of decision-making, whether they will agree to the project, or not, once they have a full and accurate understanding of the implications and effects of the project on them and their customary land. FPIC is now a key principle in international jurisprudence related to indigenous peoples.[165]

Subsidiarity is an organizing principle of decentralization under which decision-making on matters of community interest are handled by the smallest community-based, or least centralized, authority capable of addressing them effectively.

The full opposite of subsidiarity is slavery, where the chance to live by your own choices is extremely limited. One example is from Mahatma Gandhi's political movement in India when the colonial administration declared it illegal for Indians to weave cloth from the cotton they grew. The cotton was exported to the British Isles where it was dyed, woven and made into clothing that was then shipped back to India for sale. This ensured British profit while it prevented the people of India from exercising economic independence. This relationship of domination and exploitation by both state and corporate actors, established in the colonial era, remains both an overt and subtle reality that works against the realization of food security and food sovereignty in many regions of the world, including regions of affluence.

Grave limitations on self-determination are imposed when land is taken by governments and corporations for resource exploitation of one kind or another. When people are crowded into slums without public services or work opportunities they have greatly reduced self-determination and little opportunity for participation in decision-making about the matters that directly affect their lives and livelihoods.

The principle of subsidiarity is anchored in the moral right of peoples to control the resources of their region, and in the political efficacy of decision-making on public interest matters by the people closest to and most affected by them.

For these reasons the principle of subsidiarity should be centrally incorporated into national and international agreements and laws. Implementing this political principle would go far to reverse the social and economic regression that smallholder farmers, indigenous peoples, and all other marginalized people suffer when their lives are circumscribed by decisions in which they had no part.

Some international agreements now require consultation with stakeholders, which means everyone who is affected by the decisions being made. For example, the ***1998 Aarhus Convention*** of the UN Economic Commission for Europe includes European countries, but is open to other countries if approved by the Meeting of the Parties (the countries that have signed on). The Convention requires public participation at three levels: access to information, participation in decision making, and access to justice in environmental matters. Governments are required to provide timely, effective reports on the potential impact of proposed activities and alternatives. An independent grievance mechanism is required and policies must be implemented without discrimination as to citizenship, nationality or domicile.[166]

The Committee on World Food Security (CWFS) responded to the crisis in food security brought on by rising food prices in 2007-2008 followed by the global economic crisis of 2009. The CWFS instituted reforms so that it could play a more vital role in the challenge of eradicating hunger. A key part of this reform was the expansion of participation by stakeholders so that all were heard in the policy debate on food and agriculture. Special effort was made to involve those representing indigenous peoples, smallholder family farmers, fisherfolk, herders, agricultural workers, food workers, landless people, the urban poor, women, youth, and consumers.[167]

Another approach to the management of common resources is by ***self-governing organizations***. One study looked at management systems that have been operating for centuries, such as irrigation systems in Spain that date back to 1435 already mentioned, and high-mountain grazing areas in Switzerland that date back to a document signed in 1483. These systems were compared with those that did not last and the following characteristics of successful governance were determined:[168]

- Clearly defined boundaries, both of the commons and of the users;
- Congruence between the local conditions and the rules that restrict time, place, technology, and quantity of use;
- Participation of those most affected in modification of the rules;
- Monitoring by the users or those accountable to the users;
- Sanctions and punishments graduated by the seriousness of the offense and the context with allowances for emergency situations;
- Rapid access to low-cost conflict resolution services;
- Right to organize not challenged by the external government; and
- Nested enterprises for commons that are parts of larger systems.

Multi-Stakeholder Platforms (MSPs) are designed to empower stakeholders and facilitate their active participation in decision making on issues that affect their lives. MSPs facilitate dialogue among groups, from local to international levels. The objective is to bring together national governments with civil society and the private sector in open sessions that foster respectfully constructive dialogue.[169]

Conversatorios of Citizen Action (CACs) focus on empowering marginalized and vulnerable groups in local communities to participate in the governance of natural resources. First stakeholders are offered capacity-building training to raise their awareness, increase knowledge, and develop non-violent communication skills. Negotiation sessions are then held in formal CAC meetings between the stakeholders and representatives of the government, institutions, or the private sector. Once formal agreements are attained, follow-up assignments are made to make sure that agreements are kept.[170]

Biocultural Community Protocols developed by indigenous peoples, and mobile or local communities, are gaining recognition as a useful means for a range of peoples and communities to articulate the way they use, manage and conserve their natural resources and traditional knowledge. They use the protocol to engage with others according to their values, and on the basis of customary, national and international rights and responsibilities. These protocols help local communities gain legal recognition for their lands, territories and the ways they want to manage their resources. Successful examples of community protocols are happening across all continents.[171]

International Organizations

Food issues are addressed by several different United Nations-based organizations, but the FAO is the United Nations organization most heavily focused on elimination of hunger, food insecurity, and malnutrition. Strategies that FAO uses include making agriculture, forestry, and fisheries more productive and sustainable; reducing rural poverty; enabling inclusive and efficient agricultural and food systems; and increasing the resilience of livelihoods to disasters. FAO maintains an extensive database that is an excellent source for any information on food issues. The FAO developed Right to Food Guidelines in 2004, which are being reviewed after ten years as this pamphlet goes to press in 2014.[172] FAO hosts the Committee on World Food Security, which is concerned with land tenure and international investment in agriculture, climate change, food price volatility and addressing food insecurity in protracted crisis.[173]

The voluntarily funded World Food Programme, part of the United Nations system, has been fighting hunger worldwide since

1961 by providing food to victims of war, civil unrest, and natural disasters. After the emergency has passed, they help communities rebuild their lives.[174] Also voluntarily funded, the United Nations International Children's Emergency Fund (UNICEF) addresses the nutrition and food needs of children.[175]

As the right to food is recognized by most countries, those international organizations dealing with human rights and humanitarian affairs address hunger and other food system issues. Under the Office of the High Commissioner for Human Rights (HCHR) the Special Rapporteur on the Right to Food monitors the situation on the right to food throughout the world, assessing general trends and making country visits.[176] The Office for the Coordination of Humanitarian Affairs (OCHA) manages the Central Emergency Response Fund and coordinates emergency responses to food insecurity around the world, including that by the World Food Programme.[177]

Special attention is paid to the food insecurity of the most vulnerable countries by the UN Office of the High Representative for the Least Developed Countries, Landlocked Developing Countries and Small Island Developing States.[178]

As this pamphlet goes to press, Sustainable Development Goals are being negotiated at the UN to take effect in 2015. They include food security, nutrition, and sustainable agriculture.[179]

With varied results, many other UN programmes address food issues. The International Fund for Agricultural Development supports smallholder farmers; the UN Development Programme, World Bank, and International Monetary Fund support development and poverty alleviation; and the World Health Organization addresses food safety.

Civil Society Building Social Movements

Civil society institutions are responding to the growing requirement for critical social and economic change, especially in regard to food insecurity. Citizen-based organizations are actively supporting trends in the deployment of renewable energy, urban agriculture, and other smart city technologies for ecological restoration and the development of equitable and ecologically sound food systems. They also support trends in the application of novel social technologies such as local currencies, generative forms of ownership, open-source innovation, and sustainable lifestyles. Three restorative trends give hope for the future:

The first is the rapid, uncoordinated flourishing of thousands of small, locally-oriented civil society organizations dedicated to the promotion of social justice and environmental integrity. Recently, some of these small organizations have come together into a few broad umbrella groups to increase their ability to project these concerns onto a global stage.

Second is the growing interest among faith communities worldwide in environmental justice and the sanctity of creation. The faithful are being called on to acknowledge the immorality of infringements on the sanctity of life. At the same time, religious denominations are reaffirming their historic role as a strong institution of civil society.

Third is the increase in initiatives directed toward the establishment of a humanely oriented economy that develops a constructive relationship with money, enabling it to move from unsustainable growth to a sufficient and equitable steady-state.[180]

La Via Campesina is an international movement that comprises about 150 local and national organizations in 70 countries from Africa, Asia, Europe and the Americas. Altogether, it represents about 200 million farmers. It is an autonomous, pluralist and multicultural movement, independent of any political, economic or other affiliation. La Via Campesina brings together millions of peasants, small and medium-size farmers, landless people, women farmers, indigenous people, migrants and agricultural workers from around the world. It defends small-scale sustainable agriculture as a way to promote social justice and dignity. It strongly opposes corporate-driven agriculture and transnational corporations that are destroying people's way of life and the natural environment.[181]

Asia and the Pacific civil society groups representing 90 organizations from 21 countries representing various major groups and stakeholders gathered in Bangkok to formulate a just and transformative development agenda towards post-2015 and beyond. On 24 August 2013 they released the ***Bangkok Declaration*** *(p. 69)*.[182]

Bangkok Declaration

Our world is currently at a crossroads. Facing multiple and interconnected crises of environment, finance, food, energy, democracy and most of all a crisis of deep inequalities, we are confronted with a challenge and an opportunity.

State policymaking over the past three decades based on a neoliberal economic model has led to wealth, power and resources accruing to a minority of the world's richest and most powerful people and corporations. Our world is now a plutocracy. This model of wealth accumulation is directly responsible for the crises we now confront.

To achieve redistributive justice and reduce economic inequalities within countries, governments must:

Develop and implement laws and policies that ensure that small farmholders, small fishingfolk, and indigenous peoples, particularly women, have access to, control over and ownership of land, fisheries, property, productive resources, information, and appropriate and environmentally sound technology.

End policies that promote land grabbing by governments, corporations, the military, and extractive industries; and implement redistributive land reform that puts ownership of land and control over natural resources back in the hands of communities, women and other marginalized groups, and strengthens agricultural productivity and livelihoods.

Peacefully resolve cross-border and internal violent conflicts which violate human rights and affect human and economic security.

Provide financial protection and subsidies to small farming communities to ensure that they can participate on an equal basis in agricultural markets.

Develop specific national-level and time-bound targets and indicators for reducing inequalities of wealth, power and resources, and promote fair asset distribution between countries, between rich and poor, between rural and urban areas, and between different social groups, including men and women;

Reform tax policies to eliminate indirect taxes, which disproportionately impact the poor; implement progressive income taxes to ensure the wealthy contribute their fair share; implement progressive capital gains taxes and financial transactions taxes to increase government revenue and reduce harmful financial speculation; and implement taxes on the inheritance of individual wealth and property.

Prioritize public financing over public private partnerships to fulfill state obligations and strengthen public institutions; set minimum tax thresholds and re-channel military spending to finance social spending; set specific budget allocation targets to guarantee the maximum allocation of resources to protecting and promoting human rights, including the right to health, education, food, and an adequate standard of living; and increase accountability for how tax revenue is spent.

One very encouraging collaborative civil society effort is in the Philippines where the second national ***Peasants-Scientists Conference*** met in 2014 with a broad alliance of farmer's organizations, NGOs, scientists, health workers, academe, and concerned individuals. Their goal is to unite peasants and scientists toward genuine rural development, resisting the role of agrochemical transnational corporations.[183]

Strategies to Prevent and Reduce Violent Conflict

Climate change is already leading to water and food shortages that will provide the potential for conflict, locally and internationally. But violent and deadly conflict is not inevitable. Measures are urgently needed now to prepare for peaceful resolution of potential conflict and for food and resource sharing. Capacity is needed at national levels to monitor the status of vulnerable communities with resources committed to relief programs already designed, as well as at international levels for timely distribution of food to areas of need.

Most of the countries at highest risk for violent conflict exacerbated by the effects of climate change do not have the capacity to respond; so international cooperation is needed to support capacity-building that addresses preparedness to support vulnerable groups, weak governance, social instability, and economic inequality.

What measures can be implemented locally, nationally and globally to keep conflict from spreading or re-emerging? In its final report, the *Carnegie Commission on Preventing Deadly Conflict* concluded, "the prevention of deadly conflict is, over the long term, too hard—intellectually, technically and politically—to be the responsibility of any single institution or government, no matter how powerful. Strengths must be pooled, burdens shared and labor divided among actors." The report concluded that preventive strategies rest on three principals:[184]

1) Early reaction to signs of trouble,
2) A comprehensive balanced approach to alleviate the pressures that trigger violent conflict, and
3) An extended effort to resolve the underlying root causes of violence.

Strategies that may be used to reduce violent conflict in the face of climate change include:

- Act politically to create good governance including stability, elimination of corruption, implementation of fair and just policies regarding land use, access to markets, food pricing, water use, and elimination of environmental degradation;
- Insist on effective governmental management of land and water resources;
- Insist that local and national governments improve, repair, and replace critical infrastructure;
- Educate ourselves and others about potential disruptions in infrastructure due to events such as hurricanes, tornadoes, and floods; and create plans personally and in groups;
- Encourage collective community action in decision making, planning and responses to climate disruptions;
- Use indigenous, local, and traditional forms of knowledge in planning and decision making;
- Create opportunities for human and economic development, especially for marginalized communities;
- Insist on non-violent, cooperative, and collaborative techniques to resolve disputes personally and collectively;
- Develop coping mechanisms, both personally and communally, for long-term events and conditions;
- Comprehend that migration and mobility are adaptation strategies in all regions of the world that experience climate variability; and
- Remember that we are our brothers' and sisters' keepers.

Summary: Many Ways to Do What Needs to Be Done

Current industrial agriculture technology is entirely dependent on fossil fuels. Now that it is clear that carbon emissions must be reduced, agriculture technology must be reformed to be more sustainable in a carbon-constrained world. Climate change will cause drought in large areas where food has been produced using only rainfall. New irrigation systems will have to be created. Agro-ecology, where ecological principles inform agricultural practices, can be a path of innovation toward increasingly sustainable food production. The establishment of local seed networks would make seeds appropriate for local conditions available at reasonable prices without the interference of transnational corporations. Concepts like free and informed consent, subsidiarity, participation of all stakeholders, governing the commons in the public interest, Conversatorios of Citizen Action, and Bio-cultural Community Protocol are all principles and practices for management of natural resources that work for the common good. UN organizations attempt to address food insecurity in different ways, and civil society is amalgamating to generate social movements. The next chapter explores the way forward. What can we do to move into a future in which food sovereignty is achieved and the potential for deadly violence over access to resources is eliminated?

Chapter Six

The Way Forward

A Quaker Response to Food, Climate and Conflict

What approach can Quakers, and all others carrying these concerns, bring to the relationship between food, climate and conflict? What are constructive responses to the challenges climate change will bring to food systems and societal relationships at local, regional and international levels? As climate change brings increased social stresses that could lead to violence, how can Quakers draw on their experience of peace-building and conflict resolution to help address the emerging inequities that will fuel destructive, violent and deadly conflict?

This pamphlet has explored how the dominant industrial food system sows the seeds of war. This is happening through a concentration of corporate control over natural resources, marginalization of local communities, increased chemical inputs that depend on fossil fuels, and volatile food prices. The Religious Society of Friends has the opportunity—and responsibility—to respond to this food-climate-conflict nexus from the history of moral concern that Quakers have developed and applied to social and economic issues for over 350 years. A clear and consistent focus on the moral context and ethical dimensions of social and economic relationships are anchored in Quaker testimonies on peace, equality, simplicity, integrity, community, and stewardship.

Peace is at the heart of the spiritual calling of the Religious Society of Friends. Quakers have long been active, both organizationally and individually, in working to resolve conflict, end violence, and forestall war and preparations for war. The rise of transnational corporations that increasingly seek commercial control over the means of life—land, water, forests, fisheries, seeds, energy, farming, and food—in order to advance their accumulation of wealth and power has added a new dimension to worldwide conflict, violence, and war.

Equality involves understanding and addressing the power relations and social marginalization involved in the corporate takeover of the food system. This is both personal and political. One might ask, "Where did my last meal come from? Do the farmers who grew my food have enough control over their land, seeds, water, crops, and markets to have an adequate income? Will they be able to respond adequately to climate change? Will farming communities worldwide be able to continue feeding themselves and their surrounding regions?"

Community means solidarity, not only with the immediate neighborhood and region, but with all those who are interlinked in the network of relationships that sustain lives, both physically and spiritually. "From how many continents did my last meal come? Did the farmer and fisher realize a fair income from their labors?"

Stewardship asks if food choices support a food system that encourages sustainable agriculture, biodiversity, and healthy ecosystems. "What is my plate telling me about my relationship with energy, with fossil fuels, with climate change and the future viability of the planet to sustain not only human civilization, but the whole commonwealth of life?"

Quaker experience with these historic testimonies tends to converge into the ethic of right relationship. These testimonies, focused through the lens of right relationship, can empower us to build equitable and peace-promoting relationships with the people who are impacted throughout our food supply chains, as well as with the environment within which we all exist. Friends can seize the opportunity to respond to the food-climate-conflict nexus equipped with the moral heritage of Quaker testimonies.

This moral perspective on public policy and the common good is not unique to the Religious Society of Friends. Many faith communities and eco-justice workers are helping to build movements based on these shared values and principles. The ethic of right relationship is now fundamental to the movement for food sovereignty, cultural resilience, and ecological integrity.

Guiding Principles

The authors of this pamphlet developed a set of guiding principles. Any action proposed should be in alignment with these principles. Nothing should violate them.

- Earth is best understood as a living entity, whose various interacting systems support the health of the planet as a whole.
- All species in the commonwealth of life on Earth have a right to equitable access to the resources of the ecosystems in which they exist.
- The human species is but one form of life among many with no special right to Earth's natural resources.
- All people have the right of access to sufficient clean water, clean air, and nutritious food that meets their caloric and dietary needs for a healthy and active life.
- No human community, government or corporate entity has the right to degrade Earth's natural resources and fragile ecosystems in a way that reduces the overall resilience and flourishing of present and future generations of human and other life.
- Carbon emissions and other wastes from human activity must be in equilibrium with the absorptive capacity of Earth.
- Indigenous peoples have gained wisdom and experience living sustainably on Earth for millennia, so their interests and perspective must be considered in any decision making, planning, and subsequent actions.
- A community has the right to give or withhold its consent to proposed projects that may affect the lands they customarily own, occupy or otherwise use (free prior and informed consent).
- Decision-making on matters of community interest are best handled by the smallest community-based, or least centralized, authority capable of addressing them effectively (subsidiarity).

Considering these guiding principles, what is the way forward? What can be done? What follows are some general policy and action recommendations. The authors of this pamphlet are all citizens of the United States, and our intended audience is in the U.S. and Canada, so these admonitions are mostly directed toward changes in our own countries. We also make some suggestions for work in an international arena.

Action on Behalf of Food Sovereignty and Peace

The theme of this pamphlet is so problematic and so vast in the complexity of its interlocking factors that it may seem discouraging to even think about what we as individuals, or as concerned groups, might do in response. But we can all act on these food issues with respect to the food system in which we participate.

Much of what must be done to advance food security and food sovereignty is in the hands of policy professionals, human rights advocates, trade negotiators, and eco-justice workers. Those people and organizations on the front lines of food sovereignty and eco-justice work will be effective in relation to the degree of support that is brought to the issues on which they are working. This means that everyone who understands what is at stake in the control of food, food security, and food sovereignty, and is concerned about the spread of violence, can bring their support to the policies and programs that are designed to forestall the growth of resource wars.

This work can be engaged in three ways: 1) working with the local and regional food systems where you live, 2) supporting eco-justice organizations and activist work for food sovereignty, and 3) engagement with the political processes of local, state or provincial, and federal government with respect to policy and programs that relate to food security and food sovereignty, both domestically and internationally. The following are some specific issues that need attention.

Ratify International Human Rights Agreements

The U.S. has refused to accept the right to food under the International Covenant on Economic, Social and Cultural Rights, one of only five countries in the world. This is unacceptable and must be changed.

- Advocate for food sovereignty as a human right.
- Initiate a conversation about this issue with letters to the editor.
- Bring the issue to the attention of your congressional representatives.
- Conduct programs in your faith community and other organizations to publicize this issue.

Cooperate in International Climate Negotiations

The U.S. and Canada have fallen behind the other industrialized countries in responding to the reality of climate change. As historically the largest source of GHGs, the United States holds the greatest responsibility for climate change. The U.S. must act quickly to curb emissions, to catch up and lead. The refusal of the U.S. and Canada to commit to international protocols has a very discouraging effect on international negotiations. If the U.S. and Canada were to play a positive role in the climate negotiations, other nations would follow.[163] Individual actions that will help to change the policies of the governments of the U.S. and Canada include:

- Write letters to the editors of local newspapers that educate the public and government representatives.
- Lobby the State Department (U.S.) and the Department of the Environment (Government of Canada) directly to change policy to one of cooperation and leadership in international negotiations.
- Organize actions on this national leadership issue in your faith community and in other organizations.

Reduce Carbon Emissions

In order to forestall even more drastic effects of climate change on food security worldwide, the U.S. and Canada must reduce their carbon emissions. There are a number of strategies that would result in significant reductions in carbon emissions and some are already in proposed legislation:

- Remove fossil fuel subsidies that are inherent in the tax structure and allowances for exploration and exploitation of fossil fuel resources.
- Institute a carbon tax appropriate to the long-term impact of carbon emissions, along with sharing of those tax revenues with low income persons who would be severely affected by rising fuel prices.
- Support the efforts of the U.S. Environmental Protection Agency/ Canadian Department of the Environment to institute controls on GHG emissions that are within their authority.
- Support the efforts of Friends Committee on National Legislation to lobby for the passage of legislation that will reduce carbon emissions in the U.S.,[164] and the efforts of Canadian

Friends Service Committee to advocate for similar reductions in Canada.[165]

- Support local and state/provincial efforts to reduce carbon emissions.

Triumph over Climate Change Deniers

In order to get climate change legislation passed, the number of Congresspersons who deny the reality of climate change must be reduced in the U.S. Congress. Support the efforts of the League of Conservation Voters and similar organizations to defeat climate change deniers running for the U.S. Congress.[166]

Deflate the Power of Corporations

- Support efforts to overturn the precedent that gives corporations the same constitutional rights as individuals.
- Support a constitutional amendment to overturn the Supreme Court decisions that lifted limits on election contributions (Citizens United and McCutcheon).
- Check to see if your state has passed a law establishing Benefit Corporations and, if not, lobby for it.
- If your state allows Benefit Corporations, publicize them with letters to the editors of your local newspaper.
- If you have connections with corporations, lobby them to change their charter to a Benefit Corporation.

Reform Agriculture

Alternative agricultural technologies, including agro-ecological methods, are available, but with the current fossil fuel subsidies, they are not yet sufficiently competitive to be widely used. The 2014 Farm Bill included some encouragement of organic farming and farmer's markets.

- Lobby for increased support of agro-ecology in the next Farm Bill. These will be such major changes that it will be necessary to lay the groundwork early.

Restore the Safety Net

It is vitally necessary to restore the social safety net and reduce inequality and poverty, especially because food costs are bound to increase, which they should for the sake of the farmer.

- Lobby your congressional representatives to raise the minimum wage.
- Lobby to maintain and strengthen SNAP (food stamps), Social Security, Medicare, Medicaid, and the Affordable Care Act.

What Can We Do?

Individual response is hard to relate to climate, food, and violence, but the following actions are ways to personally engage this complex of issues. They are also important actions on behalf of food sovereignty no matter what happens with climate change and the threat of deadly conflict.

- Don't be fearful, be concerned about the future, and remember the power of work based on Spirit-led contemplation.
- Reduce or eliminate your consumption of meat and other foods with a high carbon and ecological footprint and encourage others to do the same.
- Become and remain aware, not just of your local situation around the issues of climate change, water and food but also of the interconnectedness of people around the world who are struggling.
- Do what you can as an individual and find others with shared interests who are working toward addressing the issues in your community.
- Work with the local and regional food systems where you live, including personal, family, and neighborhood gardening; community gardens; gardening education projects; food preservation and seed saving.
- Patronize local farm markets; encourage commercial markets to purchase directly from local and regional food producers; and organize activism on their behalf.
- Educate yourself concerning these issues. Write letters to the editor. Call in to talk shows. Enlist social media.
- Scrutinize and publicly question politicians regarding their positions on these issues to help elect people to local government who will support food sovereignty work. Lobby your local, state and national politicians. Consider running for office yourself.
- Work with local and regional authorities to create a food security orientation to community development. Help initiate and implement local and regional food security programs and projects.

- Quakers can encourage their Meetings, and others can work with their faith communities, to sponsor conflict resolution training with the goal of reducing local, national, and global conflicts.
- Support the activist work of food sovereignty and eco-justice organizations in whatever way you can: financially or actively by joining actions or conducting workshops and study groups.
- Apply pressure on corporations that are creating unsustainable food production practices or harming people in various ways. Consider boycotts, shareholder resolutions, and online petitions.
- Create social structures and institutions that bypass corporate control and build local resilience, such as worker-owned, community-based enterprises.
- Encourage Quaker and other like-minded organizations to focus attention on regions where climate change is likely to lead to conflict, so that agreements can be developed instead of warfare.
- If you have the opportunity, connect with people in other regions of the world where conflict is likely. Quakers can connect with Friends in Kenya and Bolivia.
- Always remain in the awareness that you are indeed your neighbors' keeper wherever that person is; Earth is the only place we have and we want to witness to a way of life that preserves the variety and beauty that enable life as we know it and infuses our spirits with joy.

Quaker Organizations

Quakers have embodied their testimonies and concerns in organizations that work to advance justice, peace and human rights. In recent years, and especially with climate change, the ecological context of justice, peace, and human rights has come more and more into focus. The following Quaker organizations have all incorporated the eco-justice perspective into their work.

The ***Quaker United Nations Office*** (QUNO) works on conflict and cooperation around natural resource issues, promoting a positive message that cooperation over natural resources is possible, while violent conflict is never inevitable. QUNO has recently taken on the task of working with quiet diplomacy to facilitate successful negotiations under the UN Framework Convention on Climate Change.[167]

Quaker International Affairs Programme (QIAP) operated from 2001 to 2011 when the Canadian government cut off matching funds to NGOs working on human rights and justice issues and the work had to be laid down. Although short lived, QIAP is notable for having published *The Future Control of Food: A Guide to International Negotiations and Rules on Intellectual Property, Biodiversity, and Food Security,* edited by Geoff Tansey and Tasmin Rajotte. This book was immediately recognized as a major contribution to the field, and was awarded the Distinguished Book Award of 2009 by the UK Guild of Food Writers.

Friends Committee on National Legislation (FCNL) is a Quaker lobby in the public interest working on Capitol Hill in Washington since 1943. It works directly with legislators and their staff on issues related to war, militarism, and peace, poverty and inequality, energy and environment, Native American issues, U.S. foreign policy, and domestic and international security. FCNL is notable for its Peaceful Prevention of Deadly Conflict, Smart Security, and Shared Security programs. FCNL is a lead organization among faith-based lobbies in Washington.[168]

American Friends Service Committee (AFSC) is a Quaker organization that promotes lasting peace with justice, as a practical expression of faith in action. Drawing on continuing spiritual insights and working with people of many backgrounds, it nurtures the seeds of change and respect for human life that transform social relations and systems. It was founded during World War I and has since developed a wide range of relief, service, economic and social justice, and development assistance programs, both domestically and internationally.[169]

Canadian Friends Service Committee (CFSC) was founded in 1931 to address the peace and social concerns of Canadian Yearly Meeting of the Religious Society of Friends. CFSC has worked both domestically and internationally on issues of peace and justice: It has worked with refugees coming to Canada and maintains a program of support for First Nations issues and concerns. CFSC's recently formed Peace and Sustainable Communities Committee brings economics, ecology, human rights, and peace into a holistic focus.[170]

Quaker Peace and Social Witness (QPSW) of Britain Yearly Meeting of the Religious Society of Friends maintains peace education and non-violent social change training programs in the UK. QPSW focuses on economic issues and maintains a Parliamentary Liaison to express Quaker values to government. QPSW works in Europe, Africa, the Middle East, and India on conflict resolution and non-violent change, and with Quaker United Nations Office on disarmament and peace, human rights and refugees, and global economic issues.[171]

Right Sharing of World Resources (RSWR) supports grassroots income-generating projects in developing countries, led by women, using micro-credit.[172]

Quaker Earthcare Witness (QEW) has been working since 1987 to advance an environmental consciousness among members of the Religious Society of Friends in the U.S. and Canada. QEW draws on a broad base of support by including Representatives from Yearly Meetings on the Steering Committee. QEW regularly publishes *BeFriending Creation* and *Quaker Eco-Bulletin* and has produced numerous articles and publications on climate change and food issues.[173]

Quaker Institute for the Future (QIF) is dedicated to research in the manner of Friends to advance a "global future of inclusion, social justice, and ecological integrity through participatory research and discernment."[174] It is under the auspices of QIF that this pamphlet is being researched, written, and published.

Organizations Working on Food Sovereignty and Eco-Justice Issues

Bread for the World[175]
Catalyst Commons[176]
Crop Mob[177]
EcoEquity[178]
Farms to Grow[179]
Food First—Institute for Food and Development Policy[180]
Genetic Engineering Action Network[181]
Genetic Resources Action International[182]
Global Exchange[183]

Grassroots Economic Organizing[184]
Green Peace Corps[185]
Institute for Agriculture and Trade Policy[186]
International Society for Ecology and Culture[187]
La Via Campesina[188]
Multinational Exchange for Sustainable Agriculture[189]
National Campaign for Sustainable Agriculture[190]
Seed Saving Libraries[191]
Sustainable Economies Law Center[192]
U.S. Food Sovereignty Alliance[193]

Endnotes

(Complete citations in Bibliography; websites accessed 2 October 2014)

1 Quaker Institute for the Future <quakerinstitute.org>.

2 Gee, 2012, <waysofloving.blogspot.co.uk>.

3 2013 World Hunger and Poverty Facts and Statistics <worldhunger.org/articles/Learn/world%20hunger%20facts%202002.htm>.

4 Alexandratos, Nikos and Jelle Bruinsma, 2012. World Agriculture Towards 2030/2050: The 2012 Revision <fao.org/docrep/016/ap106e/ap106e.pdf>.

5 Pingali, Prabhu, 2012. Green Revolution: Impacts, Limits, and the Path Ahead. *PNAS* 109 (31): 12302-12308 <pnas.org/content/109/31/12302.full>

6 USDA, 2014 Reaching those in Need: State Supplemental Nutrition Assistance Program Participation Rates in 2011 <www.fns.usda.gov/sites/default/files/Reaching2011.pdf>.

7 Chokshi, Niraj, 2014. Why the Food Stamp Cuts in the Farm Bill Affect Only a Third of States. Washington Post February 5, 2014. <washingtonpost.com/blogs/govbeat/wp/2014/02/05/why-the-food-stamp-cuts-in-the-farm-bill-affect-only-a-third-of-states/>.

8 Cook, et al., 2013.

9 The 5th Assessment Working Group I Report from the IPCC was first released 26 September, 2013 (IPCC, 2013).

10 National Intelligence Council, 2012. *Global trends 2030: Alternative Worlds*. Washington, DC: National Intelligence Council <dni.gov/files/documents/GlobalTrends_2030.pdf>.

11 Oreskes, 2010.

12 The 5th Assessment Working Group I Report from the IPCC was first released 26 September, 2013 (IPCC, 2013). Figure 1.4 in Chapter 1 shows the observations in relationship to the trajectories in the projections.

13 Co2now.org provides current information on atmospheric carbon dioxide concentrations and IPCC answers to frequently asked questions <co2now.org/Know-GHGs/Emissions/ipcc-faq-emissions-reductions-and-atmospheric-reductions.html>.

14 Chang, 2013.

15 Hansen, et al., 2012; Easterling, *et al.*, 2000.

16 National Weather Service <nws.noaa.gov/oh/hdsc/aep_storm_analysis/8_Colorado_2013.pdf>.

17 Masters, 2013; Ghosh, 2014. Wavier jet stream may drive weathershift. BBC News 15 February 2014 <bbc.co.uk/news/science-environment-26023166>.

18 IPCC, 2014. Working Group II. Chapter 26, North America.

19 Shephard, et al, 2012.

20 Dowestt, *et al*, 1994; Levermann, 2013.

21 Church, *et al*, 2001; Church and Whilte, 2006; Church, *et al*, 2007; Church, 2011.

22 Emanuel, 2007; Renaud and Kuenzer, 2012.

23 Meadows, *et al*, 1972.

24 Randers, 2012.

25 IPCC, 2014, Working Group III on Mitigation, Summary for Policymakers.

26 Waltham, et al., 2012.

27 One Simple Idea <one-simple-idea.com/Environment1.htm>.

28 Sen and Gemenne, 2008; van Dam and Schenewerk, 1997; Nicholls et al., 2007; Huq, et al., 1999.

29 Harman, Greg, 2014. Has the Great Climate Change Migration Already Begun? *The Guardian* September 15, 2014 <theguardian.com/vital-signs/2014/sep/15/climate-change-refugees-un-storms-natural-disasters-sea-levels-environment>.

30 Fahrenthold, David A., 2010. Last House on sinking Chrespeake Bay Island Collapses. *Washington Post* October 26, 2010 <washington-post.com/wp-dyn/content/article/2010/10/24/AR2010102402996.html>.

31 Harris, Gardiner, 2014. Borrowed Time on Disappearing Land: Facing Rising Seas, Bangladesh Confronts the Consequences of Climate Change. *The New York Times* March 28, 2014 <nytimes.com/2014/03/29/world/asia/facing-rising-seas-bangladesh-confronts-the-consequences-of-climate-change.html?_r=0>.

32 United Nations Convention to Combat Desertification, 2000.

33 Nkonya, et al., 2011.

34 International Food Policy Research <ifpri.org/node/8441#fn1>.

35 Food and Agriculture Organization FAOSTAT database <faostat.fao.org>.

36 FAO <fao.org/docrep/017/i1688e/i1688e03.pdf>.

37 World Commission on Dams, 2000. *Dams and Development: A New Framework for Decision-making.* London: Earthscan Publications Ltd,. <internationalrivers.org/resources/dams-and-development-a-new-framework-for-decision-making-3939>.

38 Water Politics. Himalayan Water Security <waterpolitics.com/2013/07/23/himalayan-water-security-a-south-asian-perspective>.

39 FAO <fao.org/docrep/018/i3107e/i3107e00.htm>.

40 David R. Stewarda,1, Paul J. Brussa,2, Xiaoying Yangb, Scott A. Staggenborgc, Stephen M. Welchc, and Michael D. Apleyd, 2013. Tapping unsustainable groundwater stores for agricultural production in the High Plains Aquifer of Kansas, projections to 2110. *PNAS* 110 (37): 3277-3486 <pnas.org/content/110/37/E3477.abstract>.

41 Press Trust of India, 2013 <ndtv.com/article/cities/dalit-woman-thrashed-for-drawing-water-from-tube-well-in-upper-caste-area-414623>.

42 Thornton and Cramer, eds., 2012.

43 Thornton and Cramer, eds., 2012, and IPCC, 2014, Working Group III, Chapter 7 Food.

44 Gregory, 2009; Anon, 2006; Staley and Johnson, 2008. Malloch etal., 2006.

45 Hannukkala, et al, 2007.

46 Smit, et al., 2000.

47 Thornton, et al, 2009.

48 Thornton and Cramer, eds., 2012.

49 FAO, 2012. The State of World Fisheries and Aquaculture 2012, Food and Agriculture Organization of the United Nations, Rome. <fao.org/docrep/016/i2727e/i2727e.pdf>.

50 Brander, K.M., 2007. Global Fish Production and Climate Change. *Proc. Natl. Acad. Sci.* 104(50) 19709-14 <ncbi.nlm.nih.gov/pubmed/18077405>.

51 Subasinghe, et al., 2012.

52 John Deere <deere.com/en_US/products/equipment/planting_and_seeding_equipment/planting_and_seeding_equipment.page?>, Case IH <caseih.com/en_us/Products/Pages/products.aspx>.

53 Consultative Group on International Agricultural Research <ccafs.cgiar.org/bigfacts/global-agriculture-emissions>.

54 FAO, Liverstock's Long Shadow <fao.org/docrep/010/a0701e/a0701e00.HTM>.

55 GRAIN, 2011 <grain.org/article/entries/4357-food-and-climate-change-the-forgotten-link>; Garnett, 2008.

56 FAO, 2013. Food Wastage Footprint: Impacts on Natural Resources. Natural Resources Management Department, Food and Agriculture Organization. <fao.org/docrep/018/i3347e/i3347e.pdf>.

57 Environmental Protection Agency (EPA), 2011. Reducing Greenhouse Gas Emissions through Recycling and Composting <epa.gov/region10/pdf/climate/wccmmf/Reducing_GHGs_through_Recycling_and_Composting.pdf>; EPA <epa.gov/epawaste/conserve/composting/pubs/index.htm#ert>.

58 World Bank <siteresources.worldbank.org/INTARD/Resources/CSA_WaterManagement_portfolio.pdf>.

59 FAO <fao.org/fileadmin/templates/nr/sustainability_pathways/docs/Factsheet_SMALLHOLDERS.pdf>.

60 IPCC, 2014, Working Group III, Chapter 7 Food.

61 U.S. Supreme Court cases: Santa Clara County v. Southern Pacific Railroad (1886) and Pembina Consolidated Silver Mining Co. v. Pennsylvania (1888).

62 Shenk, 2011.

63 Tansey and Resotte, 2008.

64 The World's Water, Pacific Institute.

65 Ciscel, *et al,* 2011, pp 41-51.

66 Gretches, 1997.

67 John Light, interviewer on Moyers and Company, September 6, 2013 <billmoyers.com/2013/09/06/drought-helped-spark-syrias-civil-war-is-it-the-first-of-many-climate-wars-to-come>.

68 Barlow, 2007.

69 Gleick, 2011.

70 Smith, 2013.

71 Public Citizen <citizen.org/documents/Bolivia_(PDF).PDF>.

72 Pinsent Masons Water Yearbook 2010-2011 (12th Edition). London:Pinsent Masons JJP <wateryearbook.pinsentmasons.com/PDF/Water%20Yearbook%202010-2011.pdf>.

73 Community Environmental Legal Defense Fund <celdf.org/-1-91>.

74 Ciscel, *et al,* 2011, pp 35-40.

75 Community Environmental Legal Defense Fund <celdf.org/rights-of-nature-background>.

76 Global Alliance for the Rights of Nature <rightsofmotherearth.com/ecuador-rights-nature>.

77 Watts, Jonathan, 2013. Ecuador approves Yasuni National park oil drilling in Amazon rainforest. *The Guardian* 16 August 2013 <theguardian.com/world/2013/aug/16/ecuador-approves-yasuni-amazon-oil-drilling>.

78 Klare, 2012a.

79 Cultural Survival, 2008. Observations on the State of Indigenous Human Rights in Light of the Declaration on the Rights of Indigenous Peoples. <lib.ohchr.org/HRBodies/UPR/Documents/Session5/BZ/CS_BLZ_UPR_S5_2009_CulturalSurvival.pdf>.

80 Indigenous Environmental Network <ienearth.org/indigenous-resistance-kxl-tar-sands>.

81 Minority Rights, 2013. Belize Court Affirms Maya Land Rights But Strikes out Protective orders. <minorityrights.org/12013/pressreleases/belize-court-affirms-maya-land-rights-but-strikes-outprotective-orders.html>.

82 Convention on Biological Diversity <cbd.int/ and International Treaty on Plant Genetics <ftp.fao.org/docrep/fao/011/i0510e/i0510e.pdf>

83 Garnett, 2008.

84 Halweil, 2002.

85 Tansey and Resotte, 2008.

86 Fortune 500, 2012 Food Industry <money.cnn.com/magazines/fortune/fortune500/2012/industries/197>.

87 Gilbert, 2006. *BeFriending Creation* July-August, 2006 <quakereart-care.org/sites/quakerearthcare.org/files/bfc/BFC1904.pd>.

88 U.S. Food and Drug Administration: Guidance and Regulation <fda.gov/food/guidanceregulation/guidancedocumentsregulatoryinformation/eggs/ucm170615.htm>.

89 The Poultry Guide <thepoultryguide.com/benefits-of-free-range-chicken-farming>.

90 Union of Concerned Scientists, Our Failing Food System. <ucsusa.org/food_and_agriculture/our-failing-food-system/industrial-agriculture>.

91 Union of concerned Scientists, 2008. CAFOs Uncovered: The Untold Costs of Confined Animal Feeding Operations <ucsusa.org/food_and_agriculture/our-failing-food-system/industrial-agriculture/cafos-uncovered.html>.

92 Hubbuch, Chris. 2014. Farm Bill primer. La Crosse Tribune, March 17, 2014 <lacrossetribune.com/farm-bill-primer/article_fd38e8c7-58dd-57e2-b1ce-2b7c6370c244.html>.

93 Shahm Anup, 2008. <globalissues.org/article/758/global-food-crisis-2008 >.

94 Shah, Naup, 2013. Structural Adjustment: A Major Cause of Poverty. Global Issues, March 24, 2013. <globalissues.org/article/3/structural-adjustment-a-major-cause-of-poverty>.

95 Amalawi, Effects of Structural Adjustment Programme on Agriculture, August 2, 2014<amalawi.info/index.php/2014/08/01/effects-of-structural-adjustment-programme-on-agriculture>.

96 Tsikata, Dzodzi, Effects of Structural Adjustement on Women and the Poor. Third World Network <twnside.org.sg/title/adjus-cn.htm>.

97 Gilbert, 2008. *BeFriending Creation* July-August, 2008, p. 5 <quakerearthcare.org/sites/quakerearthcare.org/files/bfc/BFC2104.pdf>.

98 Wise, Timothy A. New Data Confirms Food Crisis Model: Triple Crisis <triplecrisis.com/new-data-confirms-food-crisis-model>.

99 Food and Water Watch, 2008, What's Behind the Global Food Crisis? Washington D. C. <seattleglobaljustice.org/wp-content/uploads/GlobalFoodCrisisFoodWaterWatch.pdf>.

100 Clinton, President Bill, October 23, 2008 <cbsnews.com/news/bill-clinton-we-blew-it-on-global-food>.

101 Universal Declaration of Human Rights, Article 25 <un.org/en/documents/udhr/index.shtml>.

102 World Food Summit, 1996, United Nations Food and Agriculture Organization, An Introduction to the Basic Concepts of Food Security <fao.org/docrep/013/al936e/al936e00.pdf>.

103 FAO, 2008 <fao.org/docrep/013/al936e/al936e00.pdf>.

104 IPCC, 2014, Working Group III, Chapter 7 Food.

105 Declaration of Nyéléni, 13 February, 2007. <nyeleni.org/spip.php?article290>.

106 UN Committee on Economic, Social and Cultural Rights in its General Comment 12 <fao.org/righttofood/right-to-food-home/en>.

107 de Schutter <srfood.org/en/right-to-food>.

108 Anderson, Molly, 2013. Beyond Food Security to Realizing Food Rights. *U.S. Journal of Rural Studies* 29: 113-122.

109 U.S. Department of Agriculture, Food and Nutrition Service <www.fns.usda.gov>.

110 Roberts and Finnegan, 2013.

111 FAO, 2012. The Voluntary Guidelines on the Responsible Governance of Tenure of Land, Forests and Fisheries in the Context of National Food Security. Committee on World Food Security, FAO. <fao.org/docrep/016/i2801e/i2801e.pdf>.

112 UN General Assembly, 28 July 2010. <un.org/News/Press/docs/2010/ga10967.doc.htm>.

113 UN Committee on Economic, Social and Cultural Rights, November 2002. General Comment No. 15 'The right to water' <ohchr.org/Documents/Issues/Water/Climate_Change_Right_Water_Sanitation.pdf>.

114 de Schutter, 2010 <srfood.org/en/right-to-food>.

115 Golay, Christophe, 2013. <ohchr.org/Documents/HRBodies/HRCouncil/WGPleasants/Golay.pdf>.

116 UN Declaration on the Rights of Indigenous Peoples, 2007 <un.org/esa/socdev/unpfii/documents/DRIPS_en.pdf>.

117 Guiding Principles on Business and Human Rights, 2011. <ohchr.org/Documents/Publications/GuidingPrinciplesBusinessHR_EN.pdf>.

118 *War's Human Cost.* UN High Commission for Refugees Global Trends Annual Report 2013 <unhcr.org/5399a14f9.html>.

119 IPCC Report, 2014, Working Group II, Chapter 12 Human Security.

120 Guzman, 2013, p. 134-5.

121 Defense Science Board, 2011. Trends and Implications of Climate Change for National and International Security. Washington DC: Department of Defense <www.acq.osd.mil/dsb/reports/ADA552760.pdf>.

122 *Ibid.*

123 Guzman, 2013, p.135-136.

124 Smith and Vivekananda, 2007.

125 Guzman, 2013, p.65-66.

126 Guzman, 2013, p. 69.

127 Rural 21, *Effect of Climate Change on the Nile River Basin* <rural21.com/uploads/media/R21_Impact_of_climate_change_on_the_Nile..._0409_01.pdf>.

128 *Ibid*, p. 169.

129 *Ibid*, p. 145-151.

130 Future Directions International, 2012. Food and Water Insecurity: International Conflict Triggers & Potential Conflict Points <www.futuredirections.org.au/files/Workshop_Report_-_Intl_Conflict_Triggers_-_May_25.pdf>.

131 Guzman, 2013, p. 53-161.

132 Symposium on "The Intersection of Climate Change and National Security in the 21st Century" <climateandsecurity.org/2013/11/18/icymi-symposium-on-the-intersection-of-climate-change-and-national-security-in-the-21st-century>.

133 Food and Agriculture Organization (FAO), 2011. The State of Food Insecurity in the World. Rome: United Nations <fao.org/docrep/014/i2330e/i2330e00.htm>.

134 Smith, Michael, 2013. South Americans Face Upheaval in Deadly Water Battles. *Bloomberg* February 12, 2013. <bloomberg.com/news/2013-02-13/south-americans-face-upheaval-in-deadly-water-battles.html>.

135 Food and Water Watch, 2008. What's Behind the Global Food Crisis? Washington D. C.: Food and Water Watch <seattleglobaljustice.org/wp-content/uploads/GlobalFoodCrisisFoodWaterWatch.pdf>.

136 Public Citizen. *Water Privatization Case Study: Cochabamba, Bolivia* <citizen.org/documents/Bolivia_%28PDF%29.PDF>.

137 Wolf, 2006.

138 Draper, 2011.

139 Heidelberg Institute for International Conflict Research, *Conflict Barometer 2013* <hiik.de/en>.

140 Guzman,2013

141 *War's Human Cost.* UN High Commission for Refugees Global Trends Annual Report 2013 <www.unhcr.org/5399a14f9.html>.

142 IPCC Report, 2014, Working Group II, Chapter 12 Human Security

143 Gillis, Justin, 2014. Panel's Warning on Climate Risk: Worst Is Yet to Come. *NY Times*, March 31, 2014 <nytimes.com/2014/04/01/science/earth/climate.html?_r=0>.

144 IPCC WGIII, chapter six, Transformative Pathways.

145 Trawick, Paul 2002. *Natural History Magazine* <naturalhistorymag.com/htmlsite/1102/1002_feature.html>.

146 Ostrom, 1990.

147 Dietz, Ostrom, and Stern, 2003.

148 Howard, 2009.

149 Altier, *et al.*, 2011.

150 Vermeulen, *et al*, 2013.

151 The Real Seed Catalogue <realseeds.co.uk> and La Via Campasina <viacampesina.org>.

152 Agro-ecology <agroecology.org>.

153 Altieri, *et al.*, 2011.

154 Francis Thicke <www.leopold.iastate.edu/news/leopold-letter/2009/fall/keys-success-radiance-dairy-using-nature-model-and-guide>.

155 International Health Concerns about TPP <aphaih.wordpress.com/2014/02/13/how-will-a-trade-agreement-the-tpp-impact-global-health/>

156 Free Speech for People <freespeechforpeople.com/node/601>.

157 Free Speech for People <freespeechforpeople.com/node/579>.

158 Benefit Corp Information Center <benefitcorp.net>.

159 Millar, 2013.

160 United Nations Global Pulse <unglobalpulse.org>.

161 International Land Coalition <landcoalition.org>.

162 Roberts and Finnegan, 2013.

163 Arnstein, 1969.

164 Roberts and Finnegan, 2013.

165 Forest Peoples Programme <forestpeoples.org/guiding-principles/free-prior-and-informed-consent-fpic>.

166 Roberts and Finnegan, 2013; Aarhus Convention, 1998.

167 FAO, 2009 <ftp://ftp.fao.org/docrep/fao/meeting/018/k7197e.pdf>.

168 Dietz, *et al.,* 2003.

169 Faysse, 2006.

170 Candelo, 2008, and Roldan, 2008.

171 <naturaljustice.org> and <community-protocols.org>

172 Right to Food <fao.org/righttofood/right-to-food-home/ar/#.U5FGpign_m0>.

173 Committee on World Food Security <fao.org/righttofood/our-work/current-projects/rtf-global-regional-level/cfs/en/#.U5FlqSgn_m0>.

174 World Food Programme <wfp.org>.

175 UNICEF United States Fund <unicefusa.org/mission/survival/nutrition>.

176 Special Rapporteur on the Right to Food <ohchr.org/EN/Issues/Food/Pages/FoodIndex.aspx>.

177 Office for the Coordination of Humanitarian Affairs <unocha.org/top-stories/all-stories/malawi-kick-starting-food-aid#>.

178 UN Office of the High Representative for the Least Developed Countries, Landlocked Developing Countries and Small Island Developing States <unohrlls.org>.

179 Sustainable Development Goals <sustainabledevelopment.un.org/index.php?menu=258>.

180 Dreby and Lumb, 2012.

181 La Via Campesina <viacampesina.org>.

182 Bangkok Declaration, 2013. *<peoplesgoals.org/bangkok-civil-society-declaration-from-inclusive-to-just-development/#sthash.71cx1XNU.dpuf>.*

183 Resist Agrochem TNCs, 2014. <facebook.com/pages/Resist-Agrochem-TNCs/159281940800359?sk=timeline>.

184 Laipdus, 1996.

185 The intended audience for this pamphlet is Quakers in the U.S. and Canada.

186 Friends Committee on National Legislation <fcnl.org/issues/energy>.

187 Canadian Friends Service Committee <quakerservice.ca>.

188 The League of Conservation Voters has been operating a successful campaign to defeat climate deniers running for the U.S. Congress. <lcv.org>.

189 QUNO <quno.org>. Heywood, 2012, shows positive examples of inter-state cooperation over water. Roberts and Finnegan, 2013, shows positive examples of resource management. Both were produced by QUNO.

190 FCNL <fcnl.org>.

191 AFSC <afsc.org>, 2005. *Putting Rights and Dignity at the Heart of the Global Economy* <afsc.org/content/dignity-and-rights-part-1>, <afsc.org/content/dignity-and-rights-part-2>, and QEB review <www.quakerearthcare.org/sites/quakerearthcare.org/files/qeb/QEB5-2-AFSCreport.pdf>.

192 CFSC <quakerservice.ca>.

193 QPSW <quaker.org.uk/faith-action>.

194 Right Sharing of World Resources <rswr.org>.

195 QEW <quakerearthcare.org>.

196 QIF <quakerinstitute.org>.

197 Bread for the World <bread.org>.

198 Catalyst Commons <catalystcommons.org>.

199 Crop Mob <cropmob.org>.

200 EcoEquity <ecoequity.org>.

201 Farms to Grow <farmstogrow.com/home0.aspx>.

202 Food First—Institute for Food and Development Policy <foodfirst.org>.

203 Genetic Engineering Action Network <geaction.org>.

204 Genetic Resources Action International <grain.org>.

205 Global Exchange <globalexchange.org>.

206 Grassroots Economic Organizing <geo.coop>.

207 Green Peace Corps <greenpeacecorps.org>.

208 Institute for Agriculture and Trade Policy <iatp.org>.

209 International Society for Ecology and Culture <localfutures.org>.

210 La Via Campesina <viacampesina.org>.

211 Multinational Exchange for Sustainable Agriculture <mesaprogram.org>.

212 National Campaign for Sustainable Agriculture <sustainableagriculture.net>.

213 Seed Saving Libraries <richmondgrowsseeds.org>.

214 Sustainable Economies Law Center <theselc.org>.

215 U.S. Food Sovereignty Alliance <usfoodsovereigntyalliance.org>.

Bibliography

(*Websites accessed 1 October 2014*)

Altieri, Miguel, Fernando Funes-Monzote, and Paulo Petersen, 2011. Agro-ecologically Efficient Agricultural Systems for Smallholder Farmers: Contributions to Food Sovereignty. *Agronomy for Sustainable Development.* France: INRA.

Anon, 2006. Bluetongue confirmed in France. *Vet. Rec.* 159:331.

Aarhus Convention, 1998. Convention on Access to Information, Public Participation in Decision-making and Access to Justice in Environmental Matters. UN Economic Commission for Europe <unece.org/fileadmin/DAM/env/pp/welcome.html>.

Arnstein, Sherry, 1969. A Ladder of Citizen Participation. *Journal of the American Institute of Planners* 35: 216–224 <lithgow-schmidt.dk/sherry-arnstein/ladder-of-citizen-participation.html>.

Baiocchi, Gianpaolo, Patrick Heller, and Marcelo K. Silva, 2011. *Bootstrapping Democracy: Transforming Local Governance and Civil Society in Brazil.* Stanford CA: Stanford University Press.

Barlow, Maude, 2007. *Blue Covenant: The Global Water Crisis and the Coming Battle for the Right to Water.* New York: The New Press.

Berry, Thomas, 1999. *The Great Work: Our Way into the Future*. New York: Bell Tower.

Bigas, Harriet, ed., 2012. *The Global Water Crisis: Addressing an Urgent Security Issue*. Papers for the InterAction Council, 2011-2012. Hamilton, Canada: UNU-INWEH <zaragoza.es/contenidos/medioambiente/onu/newsletter12/890_eng.pdf>.

Brown, Lester Russell, 2008. *Plan B 3.0: Mobilizing to Save Civilization*. New York: W.W. Norton.

Brown, Peter G., Geoffrey Garver, Keith Helmuth, Robert Howell, Steve Szeghi, 2009. *Right Relationship: Building a Whole Earth Economy.* San Francisco CA: Berrett-Koehler Publishers.

Bruinsma, Jelle, ed., 2003. *World Agriculture 2015/2030: An FAO Perspective*. Earthscan, Rome, Italy.

Candelo, C., L. Cantillo, J. Gonzalez, A.M. Roldan and N. Johnson, 2008. Empowering Communities to Co-Manage Natural Resources: Impacts of the Conversatorio de Acción Ciudadana. In *Fighting Poverty Through Sustainable Water Use*, CGIAR Challenge Programme on Water and Food.

Chang, Kenneth, 2013. Research Cites Role of Warming in Extremes. *New York Times* September 6, 2013 <nytimes.com/2013/09/06/science/earth/research-cites-role-of-warming-in-extremes.html?ref=world>.

Church, J., 2011. Revisiting the Earth's Sea-level and Energy budgets from 1961 to 2008. *Geophys. Res. Lett.* 38(18): L18601.

Church, John A. and Neil J. White, 2006. A 20th Century Acceleration in Global Sea-Level Rise. *Geophys. Res. Lett.* 33 (2006): 4.

Church, J., S. Wilson, P. Woodworth, and T. Aarup, 2007. Understanding Sea Level Rise and Variability. *Eos* 88 (4): 43–44.

Church, J. A., J.M. Gregory, P. Huybrechts, M. Kuhn, K. Lambeck, M.T. Nhuan, D. Qin, P.L. Woodworth, 2001. Changes in sea level, in *Climate Change 2001: The Scientific Basis*, edited by J. T. Houghton et al. New York: Cambridge Univ.

Ciscel, David, Barbara Day, Keith Helmuth, Sandra Lewis, and Judy Lumb. 2011. *How on Earth Do We Live Now? Natural Capital, Deep Ecology, and the Commons.* Quaker Institute for the Future Pamphlet No. 2. Caye Caulker, Belize: *Producciones de al Hamaca.*

Collier, Paul, 2010. *The Plundered Planet: Why We Must—and How We Can—Manage Nature for Global Prosperity.* New York: Oxford University Press.

Cook, John, Dana Nuccitelli, Sarah A Green, Mark Richardson, Bärbel Winkler, Rob Painting, Robert Way, Peter Jacobs, and Andrew Skuce, 2013. Quantifying the Consensus on Anthropogenic Global Warming in the scientific literature. *Env. Res. Lett.* <iopscience.iop.org/1748-9326/8/2/024024/article>.

Córdoba, D. and D. White, 2011. *Citizen Participation in Managing Water: Do Conversatorios generate collective action?* CGIAR Challenge Programme on Water and Food, Colombo, Sri Lanka p. 6, 7.

Cortright, David. 1997. *The Price of Peace: Incentives and International Conflict Prevention.* New York NY: Carnegie Commission on Preventing Deadly Conflict. Lanham MD: Rowman and Littlefield Publishers.

Cribb, Julian, 2010. *The Coming Famine: The Global Food Crisis and What We Can Do About It.* Berkeley: University of California Press.

Dietz, Thomas, Elinor Ostrom, and Paul C. Stern, 2003. The Struggle to Govern the Commons. *Science* 302: 1907–12.

Dowsett, Harry, Robert Thompson, John Barron, Thomas Cronin, Farley Fleming, Scott Ishman, Richard Poore, Debra Willard, Thomas Holtz Jr., 1994. Joint investigations of the Middle Pliocene climate I: PRISM Paleoenvironmental Reconstructions. *Global and Planetary Change* 9: 169–195.

Draper, Robert, 2011. A Rift in Paradise. *National Geographic* November 2011, 95-113.

Dreby, Ed, and Judy Lumb, 2012. *Beyond the Growth Dilemma: Toward an Ecologically Integrated Economy.* Quaker Institute for the Future Pamphlet No. 6. Caye Caulker, Belize: *Producciones de al Hamaca.*

Easterling, Gerald Meehl, Camille Parmesan, Stanley Changnon, Thomas Karl, and Linda Mearns, 2000. Climate Extremes: Observations, Modeling, and Impacts. *Science* 289, 2068. <sciencemag.org/content/289/5487/2068.short>.

Emanuel, Kerry, 2007. Environmental Factors Effecting Tropical Cyclone Power Dissipation. *Journal of Climate* 20: 5497-5509.

Faysse, Nicolas, 2006. Troubles on the Way: An Analysis of the Challenges Faced by Multi-stakeholder Platforms. *Natural Resources Forum* 90: 216-229. <wageningenportals.nl/sites/default/files/resource/nicholas_faysse_2006_analysisofchallengesfacedbymsps.pdf>.

Garnett, Tara, 2008. *Cooking up a Storm: Food, Greenhouse Gas Emissions and Our Changing Climate*. Food Climate Research Network.

Ghosh, Pallab, 2014. Wavier jet stream may drive weathershift. BBC News 15 February 2014 <bbc.co.uk/news/science-environment-26023166>.

Gleick, Peter, Lucy Allen, Juliet Cristian-Smith, Michael J. Cohen, Heather Cooley, Mattew Heberger, Jason Morrison, Meena Palaniappan, Peter Schulte, 2011. *The World's Water Volume 7*, Pacific Institute. Washington, DC: Island Press.

Goldsmith, Edward, Vananda Shiva, Martin Khor, and Helena Norberg-Hodge, 1995. *The Future of Progress: Reflections on Environment and Development*. Dartington: Green Books.

Goldsmith, Tim, 2011. "Mining Industry Trends 2011: Balancing Supply Amidst Emerging Market Growth and Government Infrastructure Demands. *PwC: Audit and Assurance, Consulting and Tax Services*. <pwc.com/gx/en/mining/publications/2011-mine-emerging-markets-and-trends.jhtml>.

Goldstein, Jeff, 1999. "Emergence as a Construct: History and Issues." *Emergence: Complexity and Organization* 1: 49-72.

Goldstein, Jeff, 2005. Emergence, Creativity and the Logic of Following and Negating. *The Innovation Journal* 10:3. <innovation.cc/peer-reviewed/goldstein3emergence_rev2v10i3a4.pdf>.

Gray, D., 2007. Rising Seas Add to Bangkok's Sinking Feeling *NBC News*. <nbcnews.com/id/21378436/ns/us_news-environment/t/rising-seas-add-bangkoks-sinking-feeling>.

Gregory, Peter J., Scott N. Johnson, Adrian C. Newton, and John S. I. Ingram, 2009. Integrating Pests and Pathogens into the Climate Change Food Security Debate. *Journal of Experimental Botany* 60 (10): 2827–2838 <http://jxb.oxfordjournals.org/content/60/10/2827.full.pdf>.

Gregosz, David, 2012. Economic megatrends up to 2020." *Analysen & Argumente* No. 106: 1-15.

Getches, David H., 1997. *Water Law in a Nutshell*, 3rd Edition. St. Paul MN: West Publishing Co.

Grinsted, A., Moore, J., and Jevrejeva, S., 2010. Reconstructing Sea Level from Paleo and Projected Temperatures 200 to 2100 Ad. *Climate Dynamics*, 34(10), 461-472.

Guzman, Andrew, 2013. *Overheated: The Human Cost of Climate Change*. Oxford: Oxford University Press.

Hajkowicz, Stefan, et al., 2012. *Our Future World: Global Megatrends That Will Change the Way We Live*. Highett, Australia: CSIRO (Commonwealth Scientific and Industrial Research Organization). <csiro.au/Portals/Partner/Futures/Our-Future-World-report.aspx>.

Halweil, Brian, 2002. *Home Grown: The Case for Local Food in a Global Market*. Worldwatch Paper 163, Worldwatch Institute <worldwatch.org/system/files/EWP163.pdf>.

Hansen, James, Makiko Sato, and Reto Ruedy, 2012. Perception of Climate Change. *Proc. Nat. Acad. Soc.* 109 (37): 14726–14727 <pubs.giss.nasa.gov/docs/2012/2012_Hansen_etal_1.pdf>.

Hansen, James, 2007, Scientific Reticence and Sea Level Rise. *Environ. Res. Lett.* 2:024002. <iopscience.iop.org/1748-9326/2/2/024002/fulltext/>.

Hawken, Paul, 2007. *Blessed Unrest: How the Largest Movement in the World Came into Being, and Why No One Saw It Coming*. New York: Viking.

Healthy Reefs Initiative, 2012. Report Card for the Mesoamerican Reef: An Evaluation of Ecosystem Health <healthyreefs.org/cms/wp-content/uploads/2012/12/2012-Report-Card.pdf>.

Heinberg, Richard, 2007. *Peak Everything: Waking up to the Century of Declines*. Gabriola, BC: New Society Publishers.

Heywood, S., 2012. *Diverting the Flow: Cooperation Over International Water Resources*, Quaker United Nations Office, Geneva <quno.org/resource/2012/10/diverting-flow-cooperation-over-international-water-resources>.

Holliday, Laura Ward, 2005. *Mediation Within: A Path to Inner Peace*. Augusta GA: Morris Publications.

Homer-Dixon, Thomas, 2013. Our Panarchic Future. *World Watch Magazine* 22(2). <worldwatch.org/node/6008> (accessed 8 Feb. 2013).

d'Hotel, Maitre E., Lemeilleur, S., E. Bienabe, 2011. *Linking smallholders to efficient markets*. Backgroud Paper. 3rd European Forum on Rural Development. Palencia, Spain. 29 March—1 April 2011.

Howard, Philip H. 2009. Visualizing Consolidation in the Global Seed Industry: 1996-2008. *Sustainability* 1(4) 1266-1287 <mdpi.com/2071-1050/1/4/1266>.

Huq, S., Z. Karim, M. Asaduzzaman, F. Mahtab, eds., 1999. *Vulnerability and Adaptation to Climate Change for Bangladest*. Dordrecht, Netherlands: Kluwer Academic Publishers.

International Labour Organization (ILO), 1989. Convention concerning Indigenous and Tribal Peoples in Independent Countries, C169 - Indigenous and Tribal Peoples Convention. <ilo.org/dyn/normlex/en/f?p=NORMLEXPUB:12100:0::NO::P12100_ILO_CODE:C169>.

IPCC, 2007. *Climate Change 2007: Impacts, Adaptation and Vulnerability.* Fourth Assessment Report of the Intergovernmental Panel on Climate Change. M.L. Parry, O.F. Canziani, J.P. Palutikof, P.J. van der Linden and C.E. Hanson, Eds. Cambridge, UK: Cambridge University Press <ipcc.ch/publications_and_data/ar4/wg2/en/contents.html>.

IPCC, 2013. *Working Group I Contribution to the IPCC Fifth Assessment Report Climate Change 2013: The Physical Science Basis Summary for Policymakers* <climatechange2013.org>.

Jevrejeva, S., J.C. Moore, A. Grinsted, and P.L. Woodworth, 2008. Recent Global Sea Level Acceleration Started over 200 Years Ago? *Geophysical Research Letters*, 35, L08715 <onlinelibrary.wiley.com/doi/10.1029/2008GL033611/full>.

Jackson, Tim, 2009. *Prosperity without Growth: Economics for a Finite Planet*. London: Earthscan.

Johnson, V., I. Fitzpatrick, R. Floyd and A. Simms (2011) What Is the Evidence that Scarcity and Shocks in Freshwater Resources Cause Conflict instead of Promoting Collaboration? *Collaboration for Environmental Evidence* CEE review 10-010 <r4d.dfid.gov.uk/PDF/Outputs/SystematicReviews/SR10010.pdf>.

Joy, Leonard, 2011. *How Does Societal Transformation Happen?: Values Development, Collective Wisdom, and Decision Making for the Common Good*. Quaker Institute for the Future Pamphlet 4. Caye Caulker, Belize: *Producciones de la Hamaca*.

Khan, M. , 2001. *National Climate Change Adaptation Policy and Implementation Plan for Guyana*. Caribbean: Planning for Adaptation to Global Climate Change, CPACC Component 4. Georgetown, Guyana: National Ozone Action Unit of Guyana/Hydrometeorological Service.

Kinver, Mark, 2008. The Ebb and Flow of Sea Level Rise. *BBC*. January 22, 2008. <news.bbc.co.uk/2/hi/science/nature/7195752.stm>.

Klare, Michael T., 2002. *Resource Wars: The New Landscape of Global Conflict*. New York: Henry Holt.

Klare, Michael T., 2012a. *The Race for What's Left: The Global Scramble for the World's Last Resources*. New York: Metropolitan Books.

Klare, Michael T., 2012b. "The New, Golden Age of Oil That Wasn't: Forecasts of Abundance Collide with Planetary Realities," *The American Empire Project*. <aep.typepad.com/american_empire_project/2012/10/the-new-golden-age-of-oil-that-wasnt.html#more>.

Korten, David C., 2006. *The Great Turning: From Empire to Earth Community*. San Francisco, CA: Berrett-Koehler.

KPMG International Climate Change and Sustainability Services, 2012. *Expecting the unexpected*. Zug, Switzerland: KPMG International Climate Change and Sustainability Services <kpmg.com/Global/en/IssuesAndInsights/ArticlesPublications/Pages/building-business-value.aspx>.

Krugman, Paul R., 2003. *The Great Unraveling: Losing Our Way in the New Century*. New York: W.W. Norton.

Laipdus, Gail W., 1996. *Preventing Deadly Conflict: Strategies and Institutions*. Proceedings of a Conference in Moscow, Russian Federation. Carnegie Commission on Preventing Deadly Conflict <carnegie.org/fileadmin/Media/Publications/PDF/Preventing%20 Deadly%20Conflict%20Strategies%20&%20Institutions.pdf>.

Lappé, Anna, 2010. *Diet for a Hot Planet: The Climate Crisis at the End of Your Fork and What You Can Do About It*. New York: Bloomsbury.

Lappé, Francis Moore, 2011. *EcoMind: Changing the Way We Think to Create the World We Want*. New York: Nation Books.

Lappé, Francis Moore, 2013. Beyond the Scarcity Scare: Reframing the Discourse of Hunger with an Eco-Mind. *Journal of Peasant Studies*, 40(1): 219-238.

Laszlo, Ervin., 2006 *The Chaos Point: The World at the Crossroads*. Charlottesville. VA: Hampton Roads Pub. Co.

Levermann, Anders, Peter U. Clark, Ben Marzeion, Glenn A. Milne, David Pollard, Valentina Radic, and Alexander Robinson, 2013. The Multimillennial Sea-level Commitment of Global Warming. *Proc. Natl Acad. Soc. 110* (34): 13699-13700.

Lobell, David, and Marshall Burke, 2010. *Climate Change and Food Security: Adapting Agriculture to a Warmer World*. New York NY: Springer.

Lovelock, James, 2006 *The Revenge of Gaia: Earth's Climate in Crisis and the Fate of Humanity*. New York: Basic Books.

Malloch G., F. Highet, L. Kasprowicz, J. Pickup, R. Neilson, and B. Fenton, 2006. Microsatellite marker analysis of peach-potato aphids (*Myzus persicae, Homoptera: Aphididae*) from Scottish suction traps. *Bulletin of Entomological Research* 96:573–582.

Masters, Jeff, 2013. Where's spring? Second Most Extreme March Jet Stream Pattern on Record Extends Winter. <wunderground.com/blog/JeffMasters/comment.html?entrynum=2370>.

Mazur, Laurie Ann. *A Pivotal Moment: Population, Justice, and the Environmental Challenge*. Washington, D.C.: Island Press, 2010.

McKibben, Bill, 2010. *Eaarth: Making a Life on a Tough New Planet*. New York: Times Books.

McKibben, Bill, 2012. Global Warming's Terrifying New Math. *Rolling Stone* 19 July 2012.

McKinsey & Company, 2009. *Charting Our Water Future: Economic Frameworks to Inform Decision-making*. New Delhi: 2030 Water Resources Group. <www.mckinsey.com/App_Media/Reports/Water/Charting_Our_Water_Future_Full_Report 001.pdf>.

Meadows, Donella, Dennis Meadows, Jorgen Randers and William Behrens, III, 1972. *The Limits to Growth: A Report for the Club of Rome's Project on the Predicament of Mankind*. New York: Universe Books.

Meier, Mark F., Mark B. Dyurgerov, Ursula K. Rick, Shad O'Neel, W. Tad Pfeffer, Robert S. Anderson, Suzanne P. Anderson, Andrey F. Glazovsky, 2007. Glaciers Dominate Eustatic Sea-Level Rise in the 21st Century. *Science* 317: 1064–1067.

Millar, David, 2013. The New Call for a Jubilee: How Friends Can Aid a Convergence. *Quaker Eco-Bulletin* 13:1 <quakerearthcare.org/sites/quakerearthcare.org/files/qeb/qeb-13-1-jublilee-web-final.pdf>.

Ministry of the Environment, Japan, 2005. *Study Report on Comprehensive Support Strategies for Environment and Development in the Early 21st Century: Arab Republic of Egypt*. Overseas Environmental Cooperation Center, Japan <env.go.jp/earth/coop/coop/document/c_report/egypt_h16/english/pdf/000.pdf>.

Milne, G. A., J. L. Davis, Jerry X. Mitrovica, H.-G. Scherneck, J. M. Johansson, M. Vermeer, H. Koivula, 2001. Space-Geodetic Constraints on Glacial Isostatic Adjustment in Fennoscandia. *Science* 291: 2381–2385.

Milanovic, Bronko, 2012. Global Income Inequality by the Numbers: In History and Now: An Overview. *World Bank Policy Research Working Paper No. 6259* <http://elibrary.worldbank.org/doi/pdf/10.1596/1813-9450-6259>.

Morrissey, S. K., J. F. Clark, M. W. Bennett, E. Richardson, M. Stute, 2008. Effects of Sea Level Rise on Groundwater Flow Paths in a Coastal Aquifer System. Eos Trans. AGU, 89 (23), Jt. Assem. Suppl., Abstract H41C-07.

NASA, 2008. Regional Patterns of Sea Level Change 1993-2007," NASA Earth Observatory <earthobservatory.nasa.gov/IOTD/view.php?id=8875>.

National Research Council, 2013. *Executive Summary, National Climate Assessment: Draft for Public Comment* <ncadac.globalchange.gov/download/NCAJan11-2013-publicreviewdraft-chap1-execsum.pdf>.

Nicholls, R. J., S. Hanson, C. Herweijer, N. Patmore, S. Hallegatte, J. Corfee-Morlot, J. Chateau, R. Muir-Wood, 2008. Ranking Port Cities with High Exposure and Vulnerability to Climate Extremes. *OECD Environmental Working Papers (Organisation for Economic Co-operation and Development)* <oecd-ilibrary.org/environment/ranking-port-cities-with-high-exposure-and-vulnerability-to-climate-extremes_011766488208>.

Nicholls, R.J., P.P. Wong, V.R. Burkett, J.O. Codignotto, J.E. Hay, R.F. McLean, S. Ragoonaden and C.D. Woodroffe, 2007. Coastal systems and low-lying areas. In *Climate Change 2007: Impacts, Adaptation and Vulnerability*. Fourth Assessment Report of the Intergovernmental Panel on Climate Change. M.L. Parry, O.F. Canziani, J.P. Palutikof, P.J. van der Linden and C.E. Hanson, Eds. Cambridge, UK: Cambridge University Press.

Nkonya, Ephraim, Jawoo Koo, Paswel Marenya, and Rachel Licker, 2011. *Global Food Policy Report*. International Food Policy Research Institute. <ifpri.org/gfpr/2011>.

NOAA, 1995. What is an El Niño? National Oceanic and Atmospheric Administration (NOAA) <pmel.noaa.gov/tao/elnino/el-nino-story.html>.

Norgaard, Richard B., 1994. Development Betrayed: The End of Progress and a Coevolutionary Revisioning of the Future. London: Routledge.

Oreskes, N. and E. M. Conway. 2010. *Merchants of Doubt*. New York: Bloomsbury Press.

Ostrom, Elinor, 1990. *Governing the Commons: The Evolution of Institutions for Collective Action*. New York NY: Cambridge University Press.

Palmer, M., K. Haines, S. Tett, and T. Ansell, 2007. Isolating the signal of ocean global warming. *Geophysical Research Letters*, 34, L23610.

Paska, Cleo, 2010. *Global Warring: How Environmental, Economic, and Political Crises Will Redraw the World Map*. Toronto: Key Porter Books.

Patel, Raj, 2008. *Stuffed and Starved: Markets, Power, and the Hidden Battles for the World's Food System*. New York: HarperCollins.

Patel, Raj, 2009. *The Value of Nothing: How to Reshape Market Society and Redefine Democracy*. New York: Picador.

Pfeffer, W., Harper, J., and O'Neel, S., 2008. Kinematic Constraints on Glacier Contributions to 21st-Century Sea-Level Rise. *Science*, 321(5894), 1340-1343.

Rahmstorf, Stefan, 2007. Sea-Level Rise: A Semi-Empirical Approach to Projecting Future. *Science* 315: 368–370 <sciencemag.org/content/315/5810/368.abstract>.

Randers, Jorgen, 2012. 2052—*A Global Forecast for the Next Forty Years: A Report to the Club of Rome Commemorating the 40th Anniversary of* The Limits to Growth. White River Junction VT: Chelsea Green Publishing <2052.info>.

Raskin, Paul, 2002. *Great Transition: The Promise and Lure of the Times Ahead*. Boston: Stockholm Environment Institute.

Rees, Martin J., 2003. *Our Final Hour: A Scientist's Warning How Terror, Error, and Environmental Disaster Threaten Humankind's Future in this Century on Earth and Beyond*. New York: Basic Books.

Renaud, Fabrice and C. Kuenzer, Eds., 2012. *The Mekong Delta System. Interdisciplinary Analyses of a River Delta*. New York: Environmental Science and Engineering, Springer.

Roberts, Ellie, and Lynn Finnegan, 2013. *Building Peace around Water, Land, and Food: Policy and Practice for Preventing Conflict*. Quaker United Nations Office, Geneva <quno.org/sites/default/files/resources/QUNO%20%282013%29%20Building%20peace%20around%20water%20land%20and%20food.pdf>.

Roberts, Paul, 2008. *The End of Food.* Boston: Houghton Mifflin.

Rockström, Johan, 2009. A Safe Operating Space for Humanity. *Nature* 461: 472-475.

Rodale, Maria, 2010. *Organic Manifesto: How Organic Farming Can Heal Our Planet, Feed the World, and Keep Us Safe.* New York: Rodale Books.

Roldan, A.M., 2008. *A Collective Action to Recognize Commons and to Adopt Policies at Multiple Government Levels.* World Wildlife Fund Colombia.

Romm, Joseph, 2011. Desertification: The Next Dust Bowl. *Nature* 478: 450-51.

Ruggie, John, 2011 Report of the Special Representative of the Secretary-General on the Issue of Human Rights and Transnational Corporations and Other Business Enterprises. *Human Rights Council* <ohchr.org/Documents/Issues/Business/A.HRC.17.31.Add.3.pdf>.

Sale, Peter, F., 2011. *Our Dying Planet: An Ecologist's View of the Crisis We Face.* Berkeley: University of California Press.

Saul, John Ralston, 2005. *The Collapse of Globalism, and the Reinvention of the World.* New York: The Overlook Press.

Scheierling, 2011. Adapting Soil-Water Management to Climate Change: Insights from a Portfolio Review. Climate-Smart Agriculture, World Bank.

Schneider von Deimling, T., A. Ganopolski, H. Held, S. Rahmstorf, 2006. How Cold Was the Last Glacial Maximum? *Geophys. Res. Lett.* 33:L14709.

de Schutter, Olivier, 2010. *Access to Land and the Right to Food.* Report presented to the 65th General Assembly of the United Nations (A/65/281) <srfood.org/en/access-to-land-and-the-right-to-food>.

de Schutter, Olivier, 2013. *Report of the Special Rapporteur on the Right to Food.* UN General Assembly Human Rights Council <www.srfood.org/images/stories/pdf/officialreports/20140310_finalreport_en.pdf>.

Sen, Shawn, and Francois Gemenne, 2008. Tuvalu's Environmental Migration to New Zealand. *International Conference on Environment, Forced Migration and Social Vulnerability.*

Shenk, Jon., 2011. *The Island President.* Independent Lens, Public Broadcasting System. <pbs.org/independentlens/island-president/film.html>.

Shepherd, Andrew and Duncan Wingham, 2007. Recent Sea-Level Contributions of the Antarctic and Greenland Ice Sheets. *Science* 315: 1529–32.

Shepherd, Andrew, *et al*, 2012. A Reconciled Estimate of Ice-Sheet Mass Balance. *Science* 338: 1183-9.

Smith, D. and J. Vivekananda, 2007. A Climate of Conflict: The Links between Climate Change, Peace and War. *International Alert*, Initiative for Peacebuilding–Early Warning, European Commission <gsdrc.org/go/display&type=Document&id=2976>.

Smith, D. and J. Vivekananda, 2009. Climate Change, Conflict and Fragility: Understanding the Linkages, Shaping Effective Responses. *International Alert*, Initiative for Peacebuilding–Early Warning, European Commission <ifp-ew.eu/pdf/Climate_change_conflict_and_fragility.pdf>.

Smith, Michael, 2013. South America Face Upheaval in Deadly Water Battle. *Bloomberg Business Week* <bloomberg.com/news/2013-02-13/south-americans-face-upheaval-in-deadly-water-battles.html>.

Snowden, David, 2011. Risk and Resilience: Cognitive Edge. <youtube.com/watch?v=2Hhu0ihG3kY>.

Solomon, S.*et al.*, 2007. *Climate Change 2007: The Physical Science Basis.* Fourth Assessment Report of the Intergovernmental Panel on Climate Change Cambridge, UK: Cambridge University Press.

Speth, James Gustave, 2009. *The Bridge at the Edge of the World: Capitalism, the Environment, and Crossing from Crisis to Sustainability.* New Haven, CT: Yale University Press.

Subasinghe, Rohana P., J. Richard, Arthur Devin, M. Bartley, Sena S. De Silva, Matthias Halwart, Nathanael Hishamunda, C. V. Mohan, and Patrick Sorgeloos, eds, 2012. *Farming the Waters for People and Food. Proceedings of the Global Conference on Aquaculture* 2010 <fao.org/docrep/015/i2734e/i2734e.pdf>.

Sullivan, Kathryn, Thomas R. Karl, Jessica Blunden, and Jackie Richter-Menge, 2013. State of the Climate in 2012. *Bulletin of the Amer. Met. Soc.* 91(8). National Oceanic and Atmospheric Administration <www1.ncdc.noaa.gov/pub/data/cmb/bams-sotc/2012/sotc-2012-webinar-briefing-slides.pdf>

Syvitski, J., *et al.*, 2009. Sinking deltas due to human activities. *Nature Geoscience*, 2(10), 681-686.

Tansey, Geoff, and Tasman Rajotte, eds., 2008. *The Future Control of Food: A Guide to International Negotiations and Rules on Intellectual Property, Biodiversity, and Food Security.* London: Earthscan.

Tainter, Joseph A., 1988. *The Collapse of Complex Societies.* Cambridge: Cambridge University Press.

Thornton, Philip and Laura Cramer, eds., 2012. Impacts of Climate Change on the Agricultural and Aquatic Systems and Natural Resources within CGIAR's Mandate, CCAFS Working Paper 23, CGIAR Research Program on Climate Change, Agriculture, and Food Security (CCAFS) Copenhagen, Denmark <ccafs.cgiar.org>.

Thornton P.K, J. van de Steeg, A. Notenbaert, M. Herrero, 2009. The Impacts of Climate Change on Livestock and Livestock Systems in Developing Countries: A Review of What We Know and What We Need to Know. *Agricultural Systems* 101(3):113–127 <sciencedirect.com/science/article/pii/S0308521X09000584>.

Tudge, Colin, 2007. *Feeding People is Easy.* Grosseto, Italy: Pari Publishing.

United Nations Convention to Combat Desertification, 2000. Desertification, Land Degradation, and Drought (DLDD)—Some Global Facts and Figures <unccd.int/Lists/SiteDocumentLibrary/WDCD/DLDD Facts.pdf>.

UN-Water Conference, 2011. Main outcomes of the International UN-Water Conference "Water in the Green Economy in Practice: Towards Rio+2O," Zaragoza, Spain <un.org/waterforlifedecade/green_economy_2011**>.**

Vermeulen, S.J., Challinor, A.J., Thornton, P.K., Campbell, B.M., Eriyagama, N., Vervoort, J., Kinyangi, J., Jarvis, A., Läderach, P., Ramirez-Villegas, J., Nicklin, K., Hawkins, E., and Smith, D.R. 2013. Addressing uncertainty in adaptation planning for agriculture. Proceedings of the National Academy of Sciences 110: 8357–8362 Vogel C, Smith J., 2002. The Politics of Scarcity: Conceptualising the Current Food Security Crisis in Southern Africa. *South African Journal of Science* 98:315–317.

Waelbroeck, C., L. Labeyriea, E. Michela, J.C. Duplessya, J.F. McManusc, K. Lambeckd, E. Balbona, M. Labracheriee, 2002. Sea Level and deep water temperature changes derived from benthic foraminifera isotopic records. *Quaternary Science Reviews* 21: 295–305 <sciencedirect.com/science/article/pii/S0277379101001019>.

Walthall, *et al.* 2012. *Climate Change and Agriculture in the United States: Effects and Adaptation.* Washington, DC.: USDA Technical Bulletin 1935.

Winne, Mark, 2008. *Closing the Food Gap: Resetting the Table in the Land of Plenty.* Boston: Beacon Press.

Woodbridge, Roy, 2004. *The Next World War: Tribes, Cities, Nations, and Ecological Decline.* Toronto: University of Toronto Press.

Wolf, Aaron, Annika Kramer, Alexander Carius, and Geoffrey Dabelko, 2006. Water Can Be a pathway to Peace, Not War. In *Navigating Peace.* Woodrow Wilson International Center for Scholars 1:3.

World Bank, 2010. *Global Economic Prospects 2010 Crisis, Finance, and Growth.* Washington, DC: World Bank.

World Bank, 2011. *World Development Indicators 2011.* Washington, DC: World Bank.

Contributors

Judy Lumb is the lead editor and publishing agent for the Quaker Institute for the Future pamphlet series. She publishes environmental, spiritual, and cultural books for Belize and for Quakers through *Producciones de la Hamaca* in Caye Caulker, Belize. Although she has lived in Belize since 1987, she remains a member of Atlanta Friends Meeting. She serves on the governing boards of Quaker Institute for the Future (QIF) and Belize Audubon Society, and on the editorial teams for QIF, *What Canst Thou Say?* and Quaker Earthcare Witness.

Philip C. Emmi is a co-founder and continues on the board of the Quaker Institute for the Future (QIF). He is a regular participant at QIF's Summer Research Seminars. He serves on Quaker Earthcare Witness's United Nations Working Group and its Steering Committee. He is the Intermountain Yearly Meeting representative to the Friends Committee on National Legislation (FCNL), and he and his wife Elaine were inaugural residents with FCNL's revived Friend in Washington program. He is Professor of City and Municipal Planning at the University of Utah. He advocates for a holistic view of socio-ecological systems and a respect for the integrity of social and ecological relations. He is a member of the Salt Lake City Friends Meeting.

Mary Gilbert is a member of the Steering Committee of Quaker Earthcare Witness (QEW), where she also serves on the United Nations Working Group. For the past 15 years she has been the primary representative for QEW at the United Nations, where she has worked informally with a number of groups in the NGO community. She writes regularly for the QEW publication *BeFriending Creation* about what she learns at the UN and associated venues. Her professional career has included library work, social work and research in public health. She is a member of Friends Meeting at Cambridge (Massachusetts). She has served on numerous committees in her monthly meeting and in New England Yearly Meeting. Mary will sing at the drop of a hat.

Laura Ward Holliday is a mediator, lawyer, musician, and author. She is a world traveler with an interest in global cultures. She is a community activist and has participated in major movements such as Civil Rights, Women's Rights, and for the elimination of

Apartheid in South Africa. She has served as a United Nations election observer in the 1994 South African and Mexican elections. She serves on the governing boards for Quaker Institute for the Future, Pendle Hill, American Friends Service Committee, and Friends Committee on National Legislation. She is a participant in Quaker Institute for the Future's ongoing Summer Research Seminars. She has written articles in various publications and been featured in *Friends Journal*. She is the author of *Mediation Within: A Path to Inner Peace*. She is a member of the Live Oak Friends Meeting in Houston, Texas.

Leonard Joy is a member of Strawberry Creek Friends Meeting of Pacific Yearly Meeting and was founding Clerk of the Board of Trustees of Quaker Institute for the Future. As an academic he has held posts at University College of East Africa, University of Cambridge, London School of Economics, and University of California at Berkeley, initially as an economist and finally as convener of interdisciplinary studies. He has consulted in many countries for the World Bank, the United Nations Development Program, major United Nations agencies, the U.S. Agency for International Development, and the Ford and Rockefeller Foundations in fields ranging from governance, human rights, agricultural development policy, and malnutrition. He is the author of numerous publications, including QIF pamphlet #4, *How Does Societal Transformation Happen? Values Development, Collective Wisdom, and Decision Making for the Common Good.*

Shelley Tanenbaum combines her lifelong love of nature with many years of environmental advocacy and research. She is an environmental scientist with a focus on air quality. She is a former clerk of Quaker Institute for the Future and Quaker Earthcare Witness, and has served on the board of *Earthlight* Magazine. Currently, she serves as General Secretary for Quaker Earthcare Witness. Shelley lives in the San Francisco area and is a member of Strawberry Creek Monthly Meeting.

Quaker Institute for the Future

Advancing a global future of inclusion, social justice, and ecological integrity through participatory research and discernment.

The Quaker Institute for the Future (QIF) seeks to generate systematic insight, knowledge, and wisdom that can inform public policy and enable us to treat all humans, all communities of life, and the whole Earth as manifestations of the Divine. QIF creates the opportunity for Quaker scholars and practitioners to apply the social and ecological intelligence of their disciplines within the context of Friends' testimonies and the Quaker traditions of truth seeking and public service.

The focus of the Institute's concerns include:

- Moving from economic policies and practices that undermine Earth's capacity to support life to an ecologically based economy that works for the security, vitality and resilience of human communities and the well-being of the entire commonwealth of planetary life.
- Bringing the governance of the common good into the regulation of technologies that holds us responsible for the future well-being of humanity and the Earth.
- Reducing structural violence arising from economic privilege, social exclusion, and environmental degradation through the expansion of equitable sharing, inclusion, justice, and ecosystem restoration.
- Reversing the growing segregation of people into enclaves of privilege and deprivation through public policies and public trust institutions that facilitate equity of access to the means life.
- Engaging the complexity of global interdependence and its demands on governance systems, institutional accountability, and citizen's responsibilities.
- Moving from societal norms of aggressive individualism, winner-take-all competition, and economic aggrandizement to the practices of cooperation, collaboration, commonwealth sharing, and an economy keyed to strengthening the common good.

<quakerinstitute.org>

CPSIA information can be obtained at www.ICGtesting.com
Printed in the USA
BVOW04s1249151014

370845BV00002B/6/P